THE SOUL EDGE

FROM EXTRACTIVE TO
REGENERATIVE PERFORMANCE

THE SOUL EDGE

LEADING AT THE HUMAN FRONTIER
IN THE AGE OF INTELLIGENCE

BY ERNESTO J. GÓMEZ

The Soul Edge: Leading at the Human Frontier in the Age of Intelligence

by **Ernesto J. Gómez**

Library of Congress Number: 2026911791

eBook - 978-1-959009-36-8
Paperback- 978-1-959009-37-5
Hardcover - 978-1-959009-38-2

Published in the United States by AM1 Austin, Texas

Outside the United States - Published by Aspen Mindset 1 Monterrey, Mexico

To my wife, Thelma:
without your unwavering support,
I would not have been able to accomplish what I have.

To my daughters, Andrea, Fernanda, and Cordelia:
I am the luckiest father because I have you.
You inspire me every day through the remarkable women you have become.

To my sons-in-law, Andrés, Juan Pablo, and Joaquín:
I greatly value our conversations.
You keep my thinking sharp and help me learn every day.

To my grandchildren, Raquel, Alejandro, Mariel, Mateo, Lorenza, and Joaquín: you are a constant joy, and I cannot wait to see the people you will grow up to be.

I am blessed with the family I could only have dreamed of.
You are my beacon and the reason I strive to give my best.
I think of our family as a team—we root for one another and support each other in the pursuit of our goals.

May we continue to celebrate life together
and always cherish one another's company.

"Nothing except the soul
is worthy of wonder;
for the soul, if it be great,
nothing is greater."

– Seneca

Table of Contents

Part III
Crossing the Human Frontier

Appendices
The Regenerative System

INTRODUCTION

When the Compass Fails

In a room full of alumni from a prestigious university – consultants, bankers, entrepreneurs, school founders, and CEOs – I posed a simple question:

"How many of you would call the last five years the most difficult of your professional lives – with the pandemic, supply-chain crises, and geopolitical shocks?"

About 80 percent of the room raised their hands.

Then I asked, "Looking ahead, how many believe the next five years will be easier? About the same? Or more difficult?"

When I said "more difficult," even more hands went up.

These were resilient, battle-tested leaders – and they knew the road ahead would not get easier. The mood in the room was unmistakable: *we'd better buckle up; the ride ahead will be bumpy.*

As the keynote speaker that day, the reaction surprised me – but not much. It mirrored how I felt myself. You may feel the same.

If leading in today's environment already feels complex, the work of leadership in the years ahead will be even more demanding. Step back for a moment and notice how many forces now compete for your attention – and how often the solutions required are no longer the ones that worked before.

Maybe you've noticed that the old recipes – the ones that once guaranteed results – no longer work in a changing world. You sense that your leadership must evolve, but you're not sure how.

Perhaps you've felt that unsettling gap between what your experience tells you and what the new reality demands, especially in the age of AI.

For decades, we perfected the art of **extracting performance** – more productivity, more hours, more output. That was how value was created.

But very few know how to **regenerate performance** – how to expand human capacity, renew energy, imagination, and meaning. The new reality requires this for sustainable value creation.

That is why I wrote this book.

If you look at leadership conversations today, you'll find no shortage of guidance around empathy, authenticity, courage, and humility. We are told to be more human – to connect, to listen, to foster curiosity and engagement.

All of that is necessary – but it is not enough.

These qualities help us relate to one another, but they do not necessarily help us **renew** one another. They shape behavior, but they rarely transform *being* – the deeper dimension that allows us to grow into what we are meant to become.

If we are to regenerate performance and meet the challenges of the AI era, we must go deeper.

We must move beyond surface virtues and explore what is essentially human – the inner conditions that awaken creativity, expand possibility, and give meaning to our work.

AI, for all its disruption, also offers this opportunity. It is not just another technology wave.

It is a mirror – one that forces us to rethink what it means to be human, and how leadership must expand our capacity to bring depth and coherence to what we do.

The Soul Edge introduces a new kind of leadership – one that awakens, rather than exhausts, our human nature.

I call it **Soul-Based Regenerative Leadership**.

It is built around a baseline and four steps that help you rise to your best self and help others do the same – a new map for regenerating your people so they can reach their human frontier and create enduring value.

Ultimately, it calls for a new understanding of performance. Regeneration demands a shift in how we think and lead. To improve performance, we must remove the constraints that diminish human capacity.

The framework I propose goes beyond adapting to technological disruption. It invites us to reconnect with the inner place from which regenerative performance truly flows.

I did not set out to write a book about the soul.
I set out to solve performance problems.

But in doing so, I encountered a deeper question – one that every leader must now face:

What remains uniquely human as the source of leadership and value creation?
In an era where intelligence is increasingly automated and knowledge widely accessible, the advantage is shifting inward.

The human interior has become the new performance frontier.

Not our tools.
Not our systems.
Not even our intelligence.

Our capacity for awareness, meaning, creativity, and renewal.

This is where the future of leadership will be decided – in the one place technology cannot reach:

our Soul.

CHAPTER 1

Why *The Soul Edge* Matters Now

When you first read *Soul Edge*–in the title of this book or this chapter–something probably felt off. In business, we don't talk about souls. We talk about growth, margins, and valuations. Even when the conversation turns to leadership or culture–where we explore the human side of organizations–the word *soul* still sounds too soft, too abstract, too off-limits.

Suggesting that the way forward in business might begin by addressing our people's souls feels not only unconventional but uncomfortable–especially when linked to economic value creation.

I understand.

But pause for a moment. How strange is that, really?

Think about the conversations we now have in other spaces–conversations that once would have felt equally off-limits. Every day, millions of people share their thoughts, doubts, and even emotions with artificial intelligence. We ask an app, "What do you think?" "What would you do if you were me?" We thank it, argue with it, seek its advice, and even consult it about our relationships.

And yet many leaders hesitate to have that same kind of open, soul-level dialogue with the people they lead. We avoid speaking to their imagination, their fears, their essential qualities, and their aspirations. We rarely dare to ask the most human question of all:

How can I help you become who you are meant to be?

With AI, we open up easily. With each other, we hold back.

That is the defining paradox of our time.

It reveals exactly where leadership stands today–and why it must change.

We have become more transparent with machines than with one another.

And that, precisely, is why *The Soul Edge* matters now.

The future of leadership in the AI era has less to do with how much value we can extract from people's skills–and everything to do with how deeply we are willing to speak soul to soul with them. Value will increasingly come from people's inner being–the human frontier–not just their doing.

What lies ahead is not an extension of the past, but a transition into a fundamentally different logic of leadership.

When We Say an Organization Has a Soul, What Do We Really Mean?

We sometimes describe a team or company as having a soul–but usually from a limited perspective. Conventional wisdom points to:

- **Grit** – the refusal to quit
- **Character** – moral integrity
- **Heart** – deep internal commitment
- **Unity** – belonging and cohesion
- **Energy** – passion and vitality
- **Heritage** – shared stories and identity

These are visible signs of the inner human condition. They matter–but they are reflections of something deeper. When we speak of a strong backbone or moral spine, we are pointing toward an inner source without fully naming it.

Beyond Behavior

Most leadership models stop at behavior. They measure what can be seen–performance, results, outputs–while overlooking the inner coherence that makes those outcomes possible.

In this book, *soul* is not a metaphor for morale or culture.

It refers to an inner field of coherence that aligns what people do, how they think, who they are, and how they connect to something greater.

When that alignment is strong, people and organizations thrive. When it breaks, we enter diminishing returns–more effort, less energy; more pressure, less creativity.

Modern organizations measure almost everything–productivity, engagement, efficiency–yet very little about human capacity itself. We track output to the decimal point, but rarely pause to ask what people are truly capable of becoming.

The premise of this book is simple but powerful:

There is a deeper source within us that can be accessed–and when it is, performance rises naturally. Leaders can help unlock that source.

To understand why this shift is no longer optional, we must now look at how the very foundations of performance are being reshaped.

Why It Matters Now

We are experiencing a paradigm shift in the way we conduct business. For the last 150 years, performance has been driven by a simple equation:

Performance = Control + intelligence (codified know-how) + pressure

That equation is now being fundamentally challenged by the arrival of AI.

Sooner than we think, AI, data, and algorithms will permeate nearly everything we do–both individually and organizationally.

Intelligence—the kind that can be learned, systematized, and scaled—will become widely available and will cease to be a differentiator. Control, long expressed through structured management systems, increasingly conflicts with adaptability. And pressure no longer guarantees results; instead, it generates stress and impairs judgment.

In this new era, the true competitive advantage will not be how much we automate, but how deeply we activate what makes us human.

We must move from extractive performance to regenerative performance.

This shift is not cosmetic.
It is not an upgrade to existing leadership models.
It is a crossing—a fundamental transition in how leadership creates value and how human systems sustain performance.

The next great leap in leadership will not come from mastering another technology, but from reclaiming what technology can never replicate—the qualities that make us human.

The regenerative performance equation is built on:

Performance = Human capacity + alignment + meaning

When these elements are present, we not only amplify what is genuinely human, but create coherence and transformational energy that moves the organization forward—so that people can become what they are meant to be.

In this way, the human frontier becomes a true source of value creation–an economic force in its own right.

Napoleon understood the power of engaging the human soul centuries ago. He said:

"Man can only be governed through imagination; without imagination, he is a beast... it is by speaking to the soul that man is electrified."

It was true then.
It is exponentially truer now.

The Soul-Edge Regenerative Leadership Journey

Leadership is not just about managing work.
It is about helping people rise to their best self.

That journey unfolds through four regenerative steps, each building upon the last and expanding what people and organizations can become. It begins from a baseline state of awareness–how present, open, and coherent a person is–which quietly shapes how learning happens, how beliefs form, and how potential is expressed.

1. Skills – "I can do more than I thought."

Growth begins with doing. We build competence, master a craft, and discover that effort and learning stretch our limits. This is the first awakening: competence can expand.

2. Mindset – "My beliefs shape what's possible."

Experience teaches us that success depends as much on how we think as on what we do. When we change our beliefs, we change our future. This is the shift from competence to consciousness.

3. Essence – "This is who I truly am."

After years of performance, a deeper question arises: Who am I beyond my résumé?
We begin leading from our essential qualities–service, communication, curiosity, creativity. We discover what gives our work meaning and learn to express it. This is the identity awakening.

4. Higher Field – "I'm part of something greater."

Ultimately, leadership becomes an act of participation, not possession. We sense something larger–inspiration, a conscious field, the muses–to which we can open ourselves. We tune in to a deeper source of wisdom that informs our creativity and actions.

Your evolution follows a simple path:

Do → Think → Be → Transcend

Each step feeds the next, creating a regenerative current that renews both the leader and the system they touch. Doing shapes thinking. Thinking shapes being. Being opens transcendence–and transcendence reshapes everything that follows.

It is not a ladder to climb, but a living stream–an ongoing exchange between what we create and what creates us.

This is why the model is regenerative: it continually renews human energy rather than draining it.

Why Read This Book

What the world will look like in the next few years is something none of us can fully anticipate. If you sense that this uncertainty calls for a reassessment of your leadership–because results are falling short, because human energy in your organization is being depleted, or because AI is forcing leadership to evolve at a deeper level–this book is for you.

The Soul Edge is an invitation to rediscover that inner dimension–to lead in ways that speak to the soul, not just the role; to connect meaning with performance; and to regenerate yourself and your organization from the inside out.

The next frontier of business is, paradoxically, not technological. It is profoundly human.

The real revolution won't be artificial intelligence.
It will be awakened intelligence.

From Vision to Clearing

If the human frontier is now the true source of value in the AI era, then one truth immediately follows:

We cannot reach it with the leadership models, mental habits, and performance logic we inherited.

Most organizations are still operating inside an extractive performance equation–built on control, intelligence, and pressure–while trying to step into a future that demands a regenerative logic driven by human capacity, alignment, and meaning. The result is predictable: tension, exhaustion, and diminishing returns.

Leaders do not have the luxury of stopping the system to redesign it from scratch. They must fix the plane while flying it–letting go of what no longer works while continuing to deliver results.

Before we can build the new, we must first make space for it.

That is the purpose of Part I.

To simplify and clear–to loosen the grip of the old playbook, reduce the noise of outdated assumptions, and create the conditions for a different way of leading to emerge. From that cleared space, a regenerative leadership path can take shape.

Only then can the journey toward the human frontier truly begin.

Reader Takeaways

1. **We have reached a leadership paradox**
   We speak openly to machines but avoid soul-level conversations with people–limiting both human and economic potential.

2. **Soul is the source of performance**
   Soul is not culture or morale–it is the inner coherence that makes sustainable results possible.
3. **The old performance model is breaking**
   Control, intelligence, and pressure are no longer enough. AI is forcing a fundamental shift.
4. **Regenerative leadership is the future**
   Human capacity, alignment, and meaning–not extraction–will drive value.
5. **The Soul-Edge path**
   Skills → Mindset → Essence → Higher Field creates a living current of renewal.
6. **The Human Frontier is the new economy**
   The next wave of value will come from who people become, not just what they do.

PART I

The End of the Old Playbook: Making Space for the Human Frontier

The shift from extractive performance to regenerative performance begins with a deceptively simple act: making space.

For most organizations, the challenge is timing. They cannot stop the plane to redesign it; they must fix it while flying. That means creating space to let go of what no longer works while experimenting, adapting, and learning in real time.

Before we can move into something new, we must first clear what no longer serves.

To step into the human frontier–the emerging source of organizational value–we must confront the landscape we have inherited. For decades, leadership has been layered with models, tools, trends, and prescriptions. Each promised clarity. Together, they have produced noise. Instead of empowerment and coherence, leaders now experience overload and fragmentation. The old playbook has become dense, heavy, and increasingly ineffective.

Part I is about clearing that field.

Across the next three chapters, we examine how leadership became cluttered, why efficiency and control are no longer sufficient, and what must be released to make room for regenerative performance.

This is not an exercise in destruction, but in **distillation**–stripping leadership back to what truly matters. It means letting go of excess frameworks, tools, and narratives that once worked but now dilute focus and drain energy.

It also requires a fundamental shift in how we act. The imperative of *doing more with less* has pushed organizations to the edge of exhaustion and diminishing returns. The future belongs to those who learn to stop doing what doesn't matter–and to do **more of the few things that truly make a difference**, better, with greater intention, coherence, and impact.

Finally, Part I begins to redirect our trajectory. It moves us beyond efficiency and control toward something far more powerful: **the regenerative territory**–leadership that restores human capacity instead of consuming it.

As the old playbook loosens its grip, three questions quietly rise to the surface.

How do we lead, when control and pressure are no longer sufficient?
What do we choose to focus on, when doing more no longer produces better results?
And **where** do we direct our energy, when the source of value is shifting from extraction to regeneration?

This section is about releasing what weighs us down, refocusing on what truly matters, and realigning leadership with what the future now demands.
By the end of Part I, the old playbook will have loosened its grip–and space will have opened for the journey inward, toward the human frontier where the Soul-Based Regenerative Leadership path begins.

CHAPTER 2

Leadership Is Not What It Used to Be

Not too long ago, ordering coffee was simple.
Black or with cream? Sugar or no sugar? That was it.

Today, ordering a cup of coffee has become a multidimensional experience. You walk into a café and face a menu longer than a novel–cold brew, oat, almond, pumpkin spice, matcha latte, extra shot, sugar-free vanilla foam. By the time you finish ordering, you're not even sure it's coffee anymore.

Leadership has gone through a similar transformation.
It used to be monochromatic–straightforward, almost predictable.
Now it offers endless variations, and with them, growing confusion.

Everywhere you turn, there's a new recipe for success: the leadership secrets of Steve Jobs, the Netflix culture code, purpose-driven leadership, servant leadership, authentic leadership, the Amazon Flywheel Effect, resilient leadership.

Each promises extraordinary results. Copy and paste the method, and you'll win–or so it seems. And truth be told, each sounds at least partly right.

But when everything sounds right, how do you decide what actually works?

Many leaders are left wondering what to keep, what to drop, and which ingredients belong in their style–and which feel foreign to who they are. Some even start to wonder whether they're building a coherent approach or assembling a Frankenstein.

The Old and the New

The more complex the world becomes, the more leadership must return to what is simple and true. This new era calls for an intuitive way of leading–one that restores clarity, coherence, and humanity by reconnecting us to the basics.

To do that, we must redefine the playing field itself–grounding it in essentials that simplify rather than multiply. It begins with a clean slate and an openness to rediscover leadership as if for the first time.

Imagine waking up one day and realizing you've forgotten everything you know about leadership. Where would you begin?

Perhaps you'd strip away the conventional, "proven" rules–the things you've been told to do but never truly questioned. Or maybe you'd search for what feels alive and truthful beneath all the layers and prescriptions.

You might wonder: What remains when all the popular–but not necessarily useful–buzzwords fade away? What still guides you when the rules no longer apply? What feels right when the so-called certainty disappears?

In a time when AI and automation are taking over the mechanical side of work, which parts of leadership do you think would grow in value? What would you keep to rebuild leadership from its essential elements–those that speak to our shared humanity?

We must revisit leadership and lay a new foundation that helps us simplify what leadership is all about. This is how we make space for the next human frontier: by letting go of the periphery and focusing on the core. And for that, we need a map–a way to see where we stand, how we move with intention, and how we can build from there.

The Playing Field in the Human Frontier

At its core, leadership operates on two fundamentals: style and range. Picture them as two intersecting axes that define how you lead. Once you understand these, you can navigate any situation–whether familiar or challenging–with greater awareness and agility.

Range is how you express your strengths–and balance them through awareness.

On one side of the horizontal axis lies your **Trait**: the qualities you naturally rely on. We all have our signature strengths–for instance, decisiveness, confidence, drive, boldness, or optimism.

On the opposite side lies the **Counter**: the qualities that sit on the other end of that same spectrum–
inquiry (the counterpart to decisiveness),
humility (to confidence),
patience (to drive),
prudence (to boldness),
and realism (to optimism).

Together, **Trait ↔ Counter** form your leadership range–your ability to move fluidly between strength and its mirror complement, depending on what the moment requires.

Then comes **Style**, represented by the vertical axis–the way you engage with others while pursuing results.

Some leaders **push**: they drive, demand, and direct.
Others **pull**: they connect, invite, and inspire.

Push mobilizes action through pressure and control.
Pull mobilizes energy through trust and connection.

Now imagine these two fundamentals arranged as a diamond:

- The vertical axis runs between **Push** (at the bottom) and **Pull** (at the top).
- The horizontal axis runs between **Trait** (on the left) and **Counter** (on the right).

This is **The Leadership Diamond**–the new playing field where modern leadership unfolds.
It is simple in design, yet it offers multiple variations.

The beauty of the Diamond lies in the possibilities it opens, each depending on your awareness of the moment. Leadership isn't static–it shifts with context. Understanding the model is only the beginning; true mastery comes from sensing when to lean into a trait and when to balance it with its counter.

The Leadership Diamond

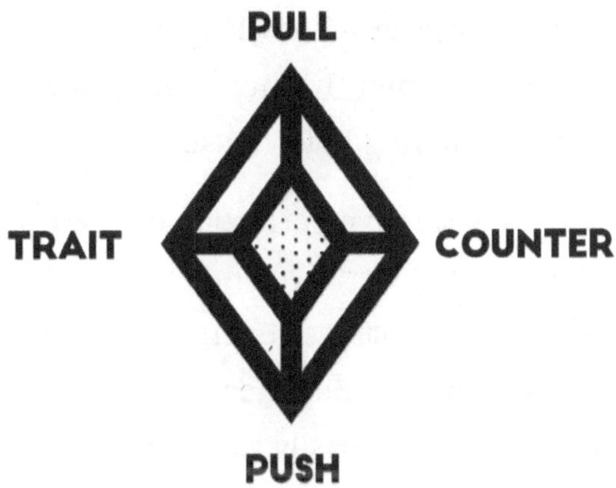

The Paradox of Traits

In today's world, one-size-fits-all leadership no longer applies. You face a challenge, so you lean into your strengths. You're decisive, confident, relentless. You take bold risks, rally the team, compete fiercely–and yet, the outcome doesn't always match the effort.

What once worked now sometimes fails, because every strength has a shadow and every situation demands a different light.

- When relentless drive turns into exhaustion, the wiser move is to pause.
- When decisiveness narrows your vision, the moment may call for questions.
- When confidence begins to distort reality, humility becomes the greater power.

That is the essence of range: acknowledging the trait while remaining open to its counter. It's not about abandoning who you are–it's about expanding how you show up.

The old playbook of extractive performance rewarded consistency. The new, regenerative one rewards discernment. Great leaders understand that a strength, left untempered, can become a weakness. Virtues and defects often share the same root; pushed too far, one can easily turn into the other.

Overdoing a virtue leads to unintended consequences, which is why flexibility becomes the safeguard. The art lies in knowing when to stay with your strength and when to move toward its counter–navigating the range that best serves the moment.

The Paradox of Style

In the past, leadership operated along a narrow path: **push to get what you want.**

The more you pushed, the more you got. The boss was never satisfied. The bar kept rising. Everything was urgent. Praise was rare. It was always: *"Good. Now do more."*

That drill-sergeant style worked in predictable, stable environments. It created urgency and delivered results–but it came at a cost: constant oversight, chronic stress, and short-term wins followed by burnout and disengagement.

That's the push approach–and yes, it's still alive in many organizations. You've heard it before: *"If you can't do it, I'll find someone who can."*

But the world has changed. Today's turbulence demands more than compliance–it demands creativity, adaptability, and initiative.

You can't push for curiosity.
You can't push for engagement.
You can't push for innovation.

The old baseline was **Push**.
The new baseline is **Pull**.

Pull leadership isn't soft or permissive. It stems from a simple belief: that people already hold immense potential–and the leader's job is to create the conditions for that potential to surface.

Does that mean every situation calls for Pull? Not necessarily. The key is judgment.
Push becomes a deliberate choice–not an unconscious habit.
Pull becomes the default–not an afterthought.

When leadership moves from Push to Pull, it stops managing behavior and begins awakening essence. It taps into the soul's energy, where authenticity and connection give rise to creative power.

The Leadership Diamond Dynamics

When we step back, Traits and Style reveal a dynamic pattern. They're not separate forces but parts of the same field–a Yin-Yang shaping how we show up and respond.

Imagine yourself moving within that diamond–up and down, left and right–adjusting to what reality requires.

When urgency calls, you may step downward into **Push**.
When trust and creativity are needed, you rise toward **Pull**.
When the team needs decisiveness, you move left into **Trait**.
When it calls for listening and reflection, you shift right into **Counter**.

The art lies in knowing when to act, how to adapt, and what energy the moment requires. This is how Traits and Style come alive in motion–flexibly, intentionally, and in service of the situation.

At any given time, you may face multiple challenges, each demanding a different combination. One moment you're leading with Pull and patience; the next, with Push and decisiveness. Most leaders have only a few well-developed strengths–five or six at best–so the key is to step back into awareness: knowing which strengths you're using, when to reach for their counter, and being flexible enough to embody your best self for the task.

The Leadership Diamond

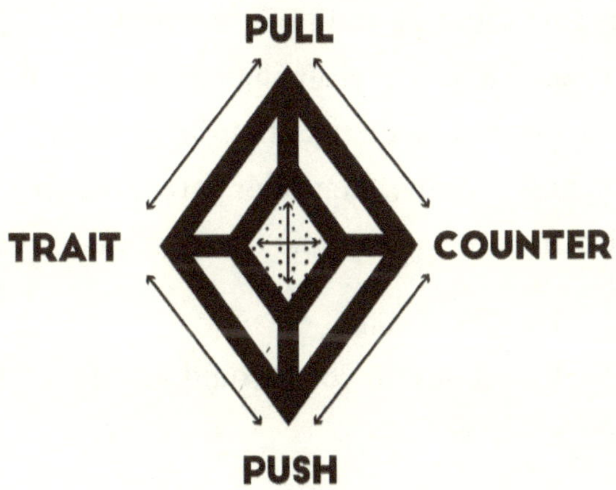

The Leadership Diamond in Action

I learned this lesson the hard way years ago when I shifted from entrepreneurial ventures to the corporate world.

To reach my goals, my instinct was to push–hard and fast. But what had once driven my success as an entrepreneur started to create resistance in a corporate setting. That experience taught me that leadership isn't about always leveraging your strongest traits, but about knowing when to balance them. Sometimes progress doesn't come from more drive, but from tempering it.

Over the years, I've seen the same principle play out again in others, in a context every leader can relate to–tight deadlines. We're constantly under pressure to deliver faster and faster. Yet sometimes, meeting the deadline carries a hidden cost: lower quality, unintended consequences, or long-term setbacks that quietly erode trust and impact.

Many leaders find themselves in that dilemma–aware that delaying a project might be wiser than hitting an arbitrary date. They can speak up and challenge their boss directly, risking confrontation and perhaps even their career. Or they can hold back, choosing instead to steer the outcome more subtly toward what's best.

If you were in that situation, what would you do–confront or steer?

This is where **The Leadership Diamond** becomes a compass.

You begin by evaluating the situation through awareness rather than reaction. You ask:
What will best serve the goal–not my ego, not my comfort, but the organization and the people involved?

From there, you sense your position within the Diamond.

If judgment calls for **Push**, you may speak up firmly, standing for your principles even at personal risk.
If the situation feels tense or fragile, you might reach for **Pull**–starting from courage's counterpoint: restraint and patience. You probe, ask questions, and guide the conversation until your boss arrives at the insight you already see.

Both paths are valid. What matters is not the action, but the awareness behind it.

Sometimes courage is loud.
Sometimes it's quiet.
Sometimes it pushes forward; sometimes it pulls from behind.

When you move through the Diamond with that awareness–between Push and Pull, between Trait and Counter–you lead with both principle and intelligence. You remain grounded in your goals, yet flexible in your approach.

That's what everyday courage truly is: not the absence of fear, but the presence of wisdom.

Awareness turns complexity into insight–and that's where simplification begins.

The Human Frontier: Simplify, Don't Multiply

Instead of endlessly adding new "features" to your leadership toolkit–like piling syrups into a coffee cup until the flavor disappears–simplify the field.

Use **The Leadership Diamond**. Return to the fundamentals.

Ask yourself:

- Should I be pushing or pulling right now?
- Am I overusing a strength that needs its counter?
- Does this moment call for direction–or space?

Breathe. Pause. Notice. Assess.

When you lead from awareness instead of formula, your actions carry coherence and power. By simplifying, you regain clarity. Leadership is no longer about holding power–it's about helping ourselves and others rise to our best selves. It must become deeper, more human, more alive.

Its purpose is not control, but the elevation of human capacity. And it begins with presence–with connecting to ourselves in a state of focused, relaxed awareness, grounded in a Pull baseline. From there, we can decide which strengths to express and which counters to invite.

That awareness is where the new human frontier begins–and where **Soul-Based Regenerative Leadership** takes root.

Reader Takeaways

1. **Simplicity is not regression–it's refinement.**
   Rediscover what is fundamental rather than adding more layers.

2. **Traits are tools, not identities.**
   Every strength has a shadow; mastery means knowing when its counter serves better.
3. **Pull before you push.**
   Begin with trust, connection, and potential; use Push only with intention.
4. **Range is the new advantage.**
   Flexing between confidence and humility, direction and reflection, force and flow defines modern leadership.
5. **Leadership's true purpose is human.**
   It's about helping others grow, awaken, and reach their best self.

CHAPTER 3

The Do More with Less Trap

"How are you holding up?"

A new colleague asked me a couple of weeks into my job at a large corporation many years ago.

"I'm doing great," I replied. "I like my job, and I'm excited to learn as much as I can from our CEO – he's known as a great leader."

"You know," he said, "he fancies himself a disciple of Jack Welch."

"No, I didn't know," I said. "How so?"

He smiled. "You'll see soon enough. By the way, you might want to pay attention to stress-related illnesses – rumor has it they're rising fast."

A few days later, I understood what he meant. My budget had been sharply reduced. Several key positions were left vacant. Everyone seemed overloaded, racing to meet goals that no longer matched their resources. The atmosphere was tense –

people were running on empty. And yes, the number of sick leaves was alarming, not just in my area but across departments.

Sound familiar?

Most organizations have lived through some version of this. It's what I now call the **pressure economy**. In this environment, the default answer to every challenge is more effort, more urgency, more output, often at the expense of the human cost.

Playing Defense

When results start to slip and profitability falls short – or when leaders want to deliver a quick, short-term gain – they tend to reach for the same lever first: **cut costs.**

It's the logical move in uncertain times. You can't blame them. When uncertainty looms, reducing feels safer than expanding. Playing defense, not offense, seems like the right move.

In corporate language, this approach comes wrapped in polished words: *rightsizing*, *streamlining operations*, or *capturing synergies*. They sound clean and rational – but are they? Behind those euphemisms are real people stretched thinner, doing the work of two or more, trying to meet goals that no longer match their resources.

More often than not, it's unsustainable – a short-term balm at best.

The message is always the same: we must "do more with less."

On paper, it makes perfect sense. Efficiency has long been a pillar of good business. The problem arises when the pursuit of efficiency loses perspective – when optimizing the parts ends up suboptimizing the whole.

Budgets get tighter. Workloads grow heavier. Energy runs thinner.

The initial gains of efficiency give way to diminishing returns, and before long, *doing more with less* turns into *doing less with less*. Productivity stalls, innovation fades, and the very people meant to drive transformation begin to burn out.

They say the poison is in the dose – and nowhere is that truer than here. The pursuit of efficiency, taken too far, begins to erode exactly what it was meant to protect: performance and value creation.

The First Casualties of the "Do More with Less" Craze

If you follow business news, you've probably noticed how often the topic of work–life balance comes up. It has become one of the defining challenges of today's workplace – and not just for seasoned professionals. Younger generations are far more vocal about the quality of their personal lives and how it shapes the way they want to work.

Sadly, burnout is no longer the exception. If you look at the data, it's fast becoming the norm. Across industries, people report feeling exhausted and disengaged while organizations scramble to respond.

Wellness programs are multiplying, mindfulness apps abound, and companies are finally talking more openly about mental health. These are encouraging signs – but they also reveal the magnitude of the problem. Leaders are beginning to realize that human sustainability isn't optional; it's a prerequisite for performance. Still, how well are we tackling the root causes?

Here's the truth: as much as organizations try to ease the pain, most are treating the symptom, not the cause.

Burnout isn't something you fix directly. It's multivariate – a symptom of how we work. The real challenge is to address what creates it: complex systems, unclear priorities, and a culture obsessed with covering every base "just in case." We've normalized overdoing and overdelivering at all costs.

Beyond the moral imperative to care for people's well-being, there are practical reasons too. Burnout and overload are quietly reshaping the talent market – both in attraction and retention. When exhaustion rises, engagement, creativity, and commitment fall. People start looking for a way out – mentally or physically – which ultimately undermines an enterprise's long-term viability.

If we truly want to make room for the **human frontier**, we need to take a hard look at how we work – and how we decide what's worth doing in the first place. We must improve both the quality and the quantity of our doing, sharpening our focus and channeling energy toward the few activities that truly move the needle.

AI Can Help - But Not Quite Yet

When the pressure to *do more with less* reaches its limit, leaders often turn to technology for relief. Artificial intelligence seems like the perfect solution – a way to automate the repetitive and give people time back for what really matters. And in time, it will deliver on that promise.

But for now, AI is making work more complicated, not less.

As companies rush to deploy new tools, workloads often spike first. Teams spend countless hours learning systems, training models, cleaning data, and adapting processes – all while keeping up with their day jobs. What was supposed to create space often compresses it further.

It's the price we pay on the road to progress: it gets worse before it gets better.

And don't expect AI to bring clarity on its own; it multiplies what already exists – clarity or chaos. In organizations where culture and leadership aren't aligned, technology amplifies the noise.

That's why many companies are realizing that success with AI depends as much on leadership capacity as on technological capability. Without focus, trust, and clear priorities, the human system becomes the bottleneck.

So yes, AI can help – but not quite yet.

Before it becomes a source of relief, it will test our ability to process complexity. The question is: will we brace for impact, or will we learn to work differently – to lead with insight, simplify, and *do an informed less?*

Doing an Informed Less: How Willing Are You to Let Go?

One of the paradoxes on the road to refocusing – escaping the *Do More with Less* trap – is that the challenge is not just strategic; it's deeply emotional.

We often believe that what we need are long strategic planning sessions to analyze the root causes of our problems. And yes, those are necessary. But the heavier burden lies elsewhere – in our **willingness to let go.**

We get attached to *doing*, often treating it as part of who we are. The busyness, the constant motion, the projects we can't seem to abandon – they make us feel useful, even indispensable. Letting go can feel like losing a piece of our identity.

It's like that old, worn pair of jeans that no longer fits but still hangs in the closet. We tell ourselves, "Maybe one day I'll lose those ten pounds and wear them again." In truth, we simply can't imagine parting with something that once defined us.

Then comes the second challenge – and this one often meets resistance from those above us: choosing what truly matters.

Letting go isn't just about stopping; it's about *selecting* – identifying the few things that genuinely move the needle and having the courage to focus there.

The path from doing the wrong things to doing the right thing well is never a single leap. As much as we'd like an instant transformation, it doesn't happen overnight. First, we must accept that moving from doing things wrong to doing the right thing requires a messy middle step: **doing the right thing poorly** – learning, adjusting, and improving until we get it right.

Nate Silver, author of *The Signal and the Noise,* captures this idea perfectly: *"The goal is to become less and less and less wrong."*

And this is where most of the pressure lies. In many workplaces, there's little patience for the "wrong." We rush for results, rewarding speed over substance. Yet progress demands the opposite: time, patience, and depth. Sometimes, **moving fast is the very reason transformation stalls.** We are human, after all – and you can't make a plant grow faster by pulling on its leaves. A tree won't bear fruit before its roots are strong.

Progress, in other words, isn't about perfection. It's about **informed subtraction** – letting go of what no longer serves so that what truly matters can take root.

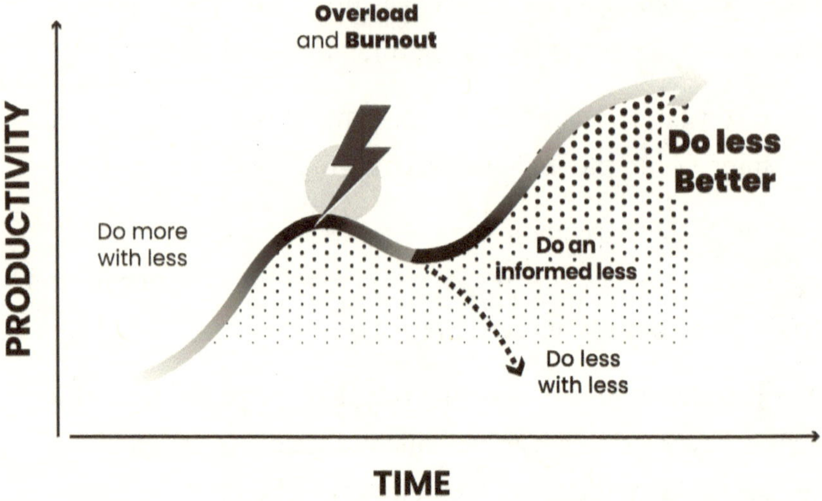

From Informed Less to Doing Less, Better

Once you've identified the few levers that truly move the needle – the activities that create disproportionate value – the next step is to **master them.** Refinement becomes your new growth strategy.

Doing less, better means channeling your energy into what matters most. When focus replaces frenzy, performance compounds, quality improves, people feel aligned, and work begins to flow again.

Consider how marathoners train. They don't run a full marathon when preparing for the race. Instead, they focus on building endurance gradually – one distance, one interval, one recovery at a time. Each session has a clear purpose. They know that trying to "do it all" would only lead to exhaustion or injury.

This is where a **regenerative flow** begins to emerge – people are clearer, energy circulates more freely, and creativity returns. Doing less, better opens the door to the human frontier. When we focus on what truly matters, we create the conditions for something deeper, which enhances human capacity: the regenerative pursuit of our best selves.

Reader Takeaways

1. **Efficiency has limits.** What begins as a smart way to stay lean can quickly turn into a self-defeating cycle when people are stretched beyond their capacity.
2. **Burnout is systemic, not personal.** It stems from how we work – unclear priorities, complexity, and constant urgency.
3. **Technology amplifies what exists.** AI won't fix broken systems; it multiplies either clarity or chaos.
4. **Letting go is strength.** Doing an informed less requires courage – the willingness to release what no longer serves and refocus on what truly matters.
5. **Mastery follows focus.** Doing less, better leads to regenerative performance – where energy, purpose, and results finally align.

CHAPTER 4

Beyond the Fog: The Quest for the Human Frontier

Every year, when I'm bombarded by reports from leadership gurus announcing the latest trends – what to focus on, what to stop doing, and what's supposedly coming next – I can't help but smile. Here we go again: it's crystal-ball season.

You've probably seen them. Just insert the current year into the following list:

- [ ] Global Human Capital Trends
- Top 5 Leadership Trends of [ ]
- [ ] Global Leaders Survey
- Leadership Trends Report [ ]
- The State of Leadership Development [ ]

These publications compile insights from CEOs, CHROs, and consultants, offering what they call *expert foresight* – tools to help leaders plan their strategies for the year ahead. It always makes me smile because those of us who have spent time formulating and implementing strategy – and especially leading transformation – know that real change takes much longer than a year.

And yet, the scene repeats itself almost everywhere:

"Wait – weren't we supposed to innovate like crazy? Now it's dropped three positions!"
"And what about resilience? We already paid the consultant for that program. Now the hot topic is critical thinking. Check if they have that course so we can switch!"
"Thank goodness we paused the agility rollout – apparently, now we need to build an AI culture!"

It's funny until it isn't. Trying to change direction on the fly usually creates confusion. Shaping culture, shifting mindsets, and building new capabilities are not short-term endeavors.

Still, I understand the appeal. Leaders crave foresight – a sense of direction and reassurance that they're on the right track and won't be left behind.

The problem is that these surveys rely on what we call the *wisdom of the crowd* – and crowds, by nature, describe what is already known. They don't reveal what's unseen or what might emerge; they reinforce what's visible. The result is a collective vision that clusters around the average, where outliers are filtered out and originality gets diluted. We start believing that if we do what everyone else is doing, we'll be fine.

There's a saying that captures the danger in that logic: "*When everybody thinks the same, nobody is thinking.*"

The paradox is that in trying to prepare for the future, we often reinforce the very patterns that keep us from seeing it. To move beyond this fog – to discern what technology will soon absorb and what will remain uniquely human – we need a clearer map. One that simplifies the landscape, helps us orient ourselves, and points toward what truly matters: the human frontier.

The Leadership Landscape

In the previous chapters, we began clearing the ground and making space to focus on what truly matters – what advances human capacity.

We stripped leadership to its essence, removing the excess baggage of complicated models and rediscovering the fundamentals. Then we refined our doing, shifting from doing more with less to doing less, better.

Now we must clarify the path forward.

To make space for the human frontier, we first need to understand where leadership stands today – and where it must go next.

Leadership today, at the organizational level, feels like an ocean of frameworks, programs, and initiatives. From agile transformations to culture resets, from well-being drives to digital upskilling, there is constant motion – but not always a clear direction. While individual leaders may work on their style and range, the system they operate in often lacks a coherent path forward. We are steering through the fog, trying to find our bearings.

To bring clarity – and a way through the maze – I have organized leadership thinking into four broad zones of focus. Together, they form what I call the **Leadership Landscape Map**.

Like any map, it simplifies reality and inevitably leaves some details behind. But it remains valuable because it helps us see patterns more clearly and understand where we truly are.

Maps can feel 30,000 feet high – just lines and boxes on a page. But this one isn't theoretical; it represents how we actually lead and where our attention goes.

As you read, think about where you and your organization spend most of your energy. Is that the best use of it? And where does your organization need to move next?

On one axis lies our focus – the inner dimension (who we are) and the outer dimension (what we do).
On the other lies our intent – the prescriptive (improving what exists) and the transformative (creating what doesn't yet exist).

The intersection of these axes yields four quadrants, each representing a distinct arena of leadership work.

The Leadership Landscape Map

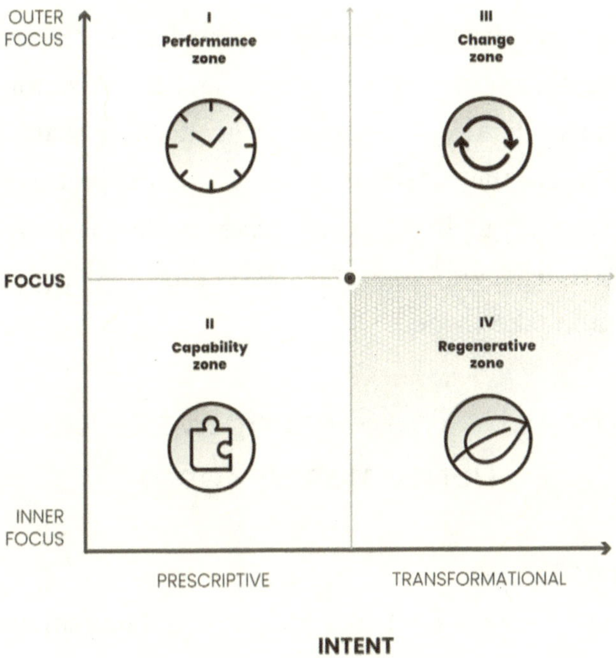

The Four Quadrants

Each quadrant represents a distinct concentration of leadership practices. They exist for a reason, and the aim here is not to criticize what's outdated or unnecessary – each plays a vital role in how organizations function and evolve.

Yet there's an essential truth to acknowledge: as AI and automation advance, the first three quadrants are reaching their natural limits. They remain indispensable, but they no longer create meaningful differentiation or the kind of leaps in growth that once set organizations apart. **This shift forces leaders to make space – to simplify, refocus, and let go of what no longer creates leverage.**

The real opportunity lies in the fourth quadrant – the space where leadership transcends efficiency and optimization to become human-centric and regenerative. It's here that leaders can fully harness the potential of the AI era, integrating technology while amplifying what makes us distinctively human. This is where regenerative performance and a new logic for value creation take shape.

Quadrant I – The Prescriptive Outer (The Performance Zone)

What it entails
This is the territory of control, efficiency, and measurable outcomes – the backbone of how most organizations operate. Leaders in this space ensure systems run smoothly and targets are met. It's about discipline, reliability, and predictable execution.

How it usually shows up
Meetings revolve around dashboards and metrics. Leaders manage through "managing by objectives," "objectives and key results," and performance scorecards to steer outcomes. Productivity reviews, cost-cutting initiatives, and process-improvement off-sites dominate the calendar. Success is measured in numbers – green indicators and year-over-year gains.

I don't know about you, but I've spent far too many Monday mornings in meetings reviewing dashboards to "evaluate performance." In my experience, a significant portion of time is consumed by executives debating the color of a dot – whether a metric is red, yellow, or green. Week after week, leaders fight for green status:

"It's not really a yellow – it's a greenish yellow."
"It's red, but we should count it as green because the competition was unusually aggressive."
"It's red, but I can turn it green if I get more budget."

The system that was designed to measure progress ends up serving as a way to save face.

Where AI is stepping in

Much of this world now runs on algorithms. AI can forecast demand, track productivity, and detect inefficiencies faster and more precisely than humans ever could. The same systems that once required managerial oversight are now capable of self-correction.

What's likely next

This quadrant will remain the foundation – it's a necessary condition, but no longer a differentiator. As efficiency becomes automated, the real advantage shifts elsewhere. Leaders must move beyond controlling results to interpreting context and anticipating shifts – adding meaning to the metrics, not merely managing them.

Quadrant II -
The Prescriptive Inner (The Capability Zone)

What it entails

This quadrant focuses on developing individuals from the inside – enhancing their skills, behaviors, and mindsets to align with the organization's goals. It's where leadership becomes about shaping people to perform better within the system that already exists.

How it usually shows up

Organizations invest heavily in training and leadership development programs. There's a wide range of initiatives – competency models, coaching certifications, and 360-degree feedback assessments. Workshops center on communication, resilience, or managing through influence. Learning paths are defined by level: "emerging leader," "new manager," "executive development."

Many companies adopt the same well-known programs, often licensed from global firms or universities. There's a sense of motion: learning portals full of micro-courses, virtual academies, and webinars on agility, empathy, or storytelling. Progress is tracked through completion rates, survey scores, and engagement dashboards.

It's all well-intentioned. I can speak from experience, having led such programs for years. But when it comes time to assess their real impact, a familiar pattern emerges: people attend, enjoy, and even quote the models. Certificates appear proudly on social media – but daily behaviors rarely shift in meaningful ways. Inevitably, you cast a wide net, and the system absorbs the effort without truly evolving.

Where AI is stepping in

AI is rapidly transforming this space. Digital learning platforms now personalize development paths in real time. Virtual coaches provide behavioral feedback and simulate difficult conversations. Analytics track engagement, sentiment, and progress continuously. Learning has become efficient and accessible – but also standardized. Everyone develops in roughly the same way, following the same playbook.

What's likely next

As automation scales learning, its impact risks becoming shallow. The next frontier in leadership development will demand greater selectivity – focusing not just on adding skills, but on expanding the inner self. Leaders will need to help people question their assumptions, look "under the hood" of their mental models, and make sense of complexity. In the Capability Zone, the challenge is no longer how much we learn, but how deeply – elevating the quality of learning as a genuine vehicle for transformation.

Quadrant III – The Transformative Outer (The Change Zone)

What it entails

This is the world of large-scale transformation – the bold attempt to reshape how an organization operates, collaborates, and behaves. It's where leadership turns its attention to redesigning structures, systems, and culture.

How it usually shows up

Companies launch sweeping programs such as *Culture 2.0*, *Digital Transformation*, or *The Future of Work*.

Consultants arrive to craft new visions, value statements, and cultural blueprints. Internal campaigns soon follow – videos, posters, town halls. Dedicated "change offices" track progress through KPIs, engagement scores, and pulse surveys. The intention is noble, but the process often feels mechanical. Culture becomes a measuring exercise rather than a renewal endeavor.

I once witnessed a company encouraging employees to "act as owners." The slogan was everywhere – and leaders were urged to reinforce it at every turn. The message was grounded in sound reasoning, but when it came to daily decisions, people hesitated. They didn't fully understand what it meant, nor did they feel empowered, trusted, or rewarded for behaving like owners. The idea was good, but it had no roots. It was a message without traction – lacking the depth of genuine commitment.

Where AI is stepping in

AI now automates parts of this process – analyzing employee comments, predicting attrition, and suggesting behavioral nudges. Dashboards visualize "culture health" in real time. It's fast, scalable, and impressive – but it still treats culture as a system to be optimized. If today's organizations are obsessed with measuring culture, AI will only multiply their capacity to do so. But is that really the way forward?

What's likely next

AI will accelerate change – but it will not deepen it. True transformation begins when people rediscover coherence – when change is embodied, not merely managed. The future will favor leaders who move from orchestrating initiatives to regenerating meaning – leaders who see culture not as something to install, but as something to embody and live through their actions.

Quadrant IV - The Transformative Inner (The Regenerative and Human-Centric Zone)

What it entails

This is the least explored – and most vital – territory. Here, leadership begins within. It's the space where awareness, aspiration, imagination, and the rise of the best self converge. Instead of managing energy, leaders learn to renew it.

How it usually shows up

Leaders in this space create moments to pause, listen, and connect. They recognize that each person follows a unique journey – and that their role is to accompany and support it. They speak to the soul, not to the role. They create environments where people can expand their awareness and uncover what is ready to emerge within them.

Where AI stops short

AI can simulate empathy, but it cannot feel it. It can analyze tone, but not intention. Regeneration lies beyond code – it belongs to the realm of creativity, intuition, and inner alignment. This is a human space, where AI remains on the periphery.

What's likely next

As automation assumes control of the outer layers of performance, the true value of leadership shifts inward. The next era will belong to those who can restore trust, coherence, and vitality – in themselves, their teams, and their organizations. This is the zone where Soul-Based Regenerative Leadership lives: leadership that regenerates the source, not just the results.

The Quadrants in Motion

The first three quadrants are not obsolete – they serve as the scaffolding on which organizations grow and evolve. They form the foundation on which leadership has long been built.

But as AI perfects their logic, leaders must elevate their aim – redirecting it toward what only humans can discern and cultivate.

Performance, capability, and change remain essential, yet their center of gravity must shift. The key is to design systems that allow AI to handle the procedural and repetitive – so that leadership can focus on what is profoundly human. This is the path to enhancing humanity and enabling true transformation.

The Human Frontier Awaits

In a world obsessed with optimization, it's easy to lose sight of the human dimension – the very force that releases energy, creativity, and ingenuity during times of transformation. This is the missing piece of the puzzle, the one you won't find in annual leadership trend reports that tell us what to fix next.

The future will belong to those who nurture the human frontier.

This is where leadership must go next – toward a deeper understanding of what renews us, what inspires trust, and what gives our work meaning. The challenge is not to do more or learn faster. Doing for the sake of doing – or learning for the sake of learning – will eventually hit a wall.

We must instead reconnect with what is real, distinctively human, and alive. The path forward begins by re-centering leadership on what no machine can replicate: our consciousness, our creativity, and our capacity for renewal.

In the chapters ahead, we will explore how to do this through the **Soul-Based Regenerative Leadership Model (SBRL)** – a practical framework that helps leaders move from extractive performance to regenerative performance, develop human capacity, and rise to our best selves.

Reader Takeaways

1. **Foresight from the crowd is hindsight in disguise.** Most "trend" reports recycle what is already visible; originality emerges from those willing to see beyond the average.
2. **Complexity calls for simplicity.** Clearing space – reducing models, metrics, and noise – creates the clarity leaders need to move forward.
3. **AI will master execution, not meaning.** Efficiency, prediction, and optimization will soon be automated; the human advantage will lie in sense-making and reconnecting with the inner source.
4. **Inner work becomes the differentiator.** Awareness, coherence, and renewal are the true sources of regenerative performance in the AI era.
5. **The future is regenerative.** As AI takes care of the procedural, leaders must design systems that amplify what is creative, conscious, and alive – the essence of the human frontier.

PART II

The New Core: Soul-Based Regenerative Leadership

In Part I, we cleared the space.
We simplified our approach to leadership, refined our doing, and clarified our direction.

We questioned the assumptions of the old playbook, recognizing that the performance equation that once delivered results is collapsing under complexity, acceleration, and the arrival of AI–and that navigating today's landscape requires a return to what is essential.

What we uncovered was simple, yet profound:
to lead effectively in the AI era, we must move toward a regenerative, human-centric zone.

Part II is about what comes next.

It marks the shift from clearing to discovering–from removing noise to revealing a coherent, step-by-step path. It explores how we place the soul at the center and rise to our best selves, not by adding complexity, but by expanding what we are capable of becoming. The aim is to amplify our own–and others'–human capacity.

This is a journey inward and upward:
from skills to mindsets,
from the soul to a higher field of creativity and resourcefulness.

Here, we introduce the Soul-Based Regenerative Leadership Model (SBRL)–a four-step framework designed to help leaders regenerate themselves and their teams from the inside out. It reconnects the full arc of leadership: doing, thinking, being, and transcending.

We will explore:

1. **Mastering the Tools (Skills)** – Doing with clarity.
2. **Expanding the Horizon (Mindset)** – Thinking differently.
3. **Discovering the Essence (Soul Signature)** – Being anchored in our core.
4. **Opening to the Higher Field (Receiving)** – Receiving insight, creativity, and inspiration.

Together, these four movements form a regenerative path–a living cycle that allows leaders to renew their energy, expand their capacity, and elevate the people around them.

And this journey begins with a fundamental capability–the one that shapes the quality of every step that follows:

Awareness.

In Chapter 5, we turn to awareness–the red thread of transformation–and explore how mastering this inner vantage point becomes the foundation for leading and living from our best selves.

CHAPTER 5

The Red Thread of Awareness

"You're a traitor," my boss said.

Then came silence.

We were walking back to our office, down the long corridors of the largest media consortium in Mexico. We had just left the office of the Executive Vice President–his boss–and what should have been a short ten-minute walk felt like an eternity. He didn't utter another word.

The story had begun a few days earlier. My boss was on vacation on a remote Polynesian island when I received a call that would test every one of my professional capacities. I was a young manager leading social studies at the country's largest television network. Late that Thursday, I was informed that a confidential survey had been requested by none other than the President of Mexico.

My boss, thousands of miles away, called and said: "You're in charge. There are no limits on expenses. Just do it."

The stakes were enormous. The President had been gifted a large ranch by a major company, and public opinion was divided. He wanted to know how the population felt–discreetly, without involving the government–so he asked our Executive Vice President to handle it personally.

Those were pre-digital days. I had to design the sample, write the questions, run the fieldwork, analyze the data, and build presentation boards–by hand. For three sleepless nights, my team and I worked relentlessly. By Sunday evening, the boards and my notes were ready. The Executive Vice President would present them to the President first thing Monday morning.

By noon that same day, my boss was back in Mexico. We were scheduled to meet with his boss later to hear the outcome. Before that meeting, he called me into his office.

He wanted to review the study.

Within minutes, he was tearing it apart.
He criticized the design, the questions, the findings. He said he regretted hiring me and called it the worst work he had ever seen.

I remember feeling the floor disappear beneath my feet. I thought I was about to be fired. Still, I stayed composed. There was nothing to do but face it.

As we walked the corridors toward his boss's office, he was still ranting. Then we entered the Executive Vice President's office.

The first thing I saw were my boards laid out on a couch and the file with my notes.

My boss began to speak–something halfway between an apology and an explanation–but his boss interrupted him.

"What are you talking about?" he said. "The survey was excellent! The President was very pleased. He personally congratulated me. The company scored a major point–we've earned his trust."

He turned toward me.
"This fine young man pulled it off in record time." He shook my hand and said, "Congratulations!"

As he continued to praise the work, my boss stood there, silent, forcing a half-smile and avoiding eye contact.

We left the meeting and walked back along the same long corridor.
That's when he said, "You're a traitor."
Those were his only words. He didn't say anything else.

He slammed his office door, and we didn't speak for days. Eventually, things returned to normal.

I never knew why he called me a traitor. Perhaps he thought I'd known the Executive Vice President was pleased and had let him humiliate himself. Maybe it was pride. It doesn't matter.

What mattered–and what stayed with me ever since–was what happened inside me that day.

Within ninety minutes, I experienced the lowest and highest moments of my professional life. One moment, I was worthless; the next, I was celebrated–king of the hill. Yet, in both moments, I was the same person. The insight I carried from then on was that, in the worst and in the best, nothing in me had changed–only what the outside world was telling me. My awareness of myself was the same.

That realization became my red thread in life: awareness is my foundation. What happened to me had nothing to do with who I truly was. I could watch myself as if I were a character in a movie, weathering a storm, and even in the tension of the moment, I was able to return to my center. I was both the actor and the spectator of the situation. I felt grateful for the outcome, but even in the moments of blame, my identity remained untouched.

When we are aware of who we are, the noise around us loses its power. Praise and blame become passing weather. What remains is the soul from which we can act with integrity, clarity, and resilience.

The Two Stages of Awareness

Through reflection and experience, I've come to see that awareness unfolds in two stages.

1. Letting reality in – surrendering to what is.
Accepting what the moment brings, even when it is not what we expected or wished for.

2. Stepping into the vantage point of the observer.
From here, we see ourselves and the situation more clearly and choose the response that truly serves us.

The following sections explore each stage.

Letting Reality In - Surrender

I was raised to believe we must always keep our guard up. Whatever happens, you respond swiftly and decisively. Otherwise, you become prey. The mindset was simple: read the situation fast, strike with your best shot, never show weakness. That's how you survive, escalate, and succeed.

So when I began exploring awareness and encountered the Zen and Taoist idea that the first step is *surrender*, my reaction was immediate:

What? Never.

Surrender felt forbidden – passive, vulnerable, incompatible with everything I'd been taught.

But over time, I discovered that, as counterintuitive as it sounds, the concept is profoundly true. The word may feel uncomfortable, but the principle is powerful.

Surrender does not mean giving up.
It means accepting what is happening – fully, without distortion – before deciding how to act.

When we don't accept reality, we filter it through denial, ego, fear, or wishful thinking. And acting from distortion almost always leads us in the wrong direction.

Life presents us all kinds of setbacks:

- You get passed over for a promotion.
- Your home leaks during a storm.
- Your pet disappears.
- Your relationship falters.
- Your job drains you.
- Your finances get tight.

When something like this happens, we tend to deny it or to avoid it – hoping it'll fix itself if we leave it alone long enough.

But denial or avoidance is not awareness.

Surrender means lowering your guard just long enough to let the truth in.

It's the moment you stop fighting reality, stop resisting your emotions, and stop forcing the world to match your expectations.

You're not giving up – you're simply seeing clearly.

And clarity is what allows right action.

When we breathe, soften, and accept, the mind becomes still enough to see the situation without the storm. It's the eye of the hurricane – a place of calm inside the chaos. From that center, we don't get swept away. We respond with presence rather than react from panic. We let our soul take the wheel.

Surrender is not passivity; it is clarity.
It creates the internal space from which wisdom emerges. And from that space, action becomes more precise, grounded, and regenerative.

This is where awareness begins – not in control, but in acceptance.

The Vantage Point of the Observer

Once we surrender, we are ready to take perspective. Awareness deepens when we access the inner vantage point of the observer – what Anthony de Mello called the witnessing self.

Try this simple exercise:

Take a sheet of paper and briefly describe yourself – your strengths, your challenges, the qualities that come to mind. Write whatever appears: parent, businessperson, curious, impatient, generous, disciplined, musician, early riser, pragmatic, creative, friendly.

Then notice what happens.

As you write, you are both:

- **the me** – the person with traits and behaviors
- **and the I** – the one who observes them

You suddenly become twofold.
The "I" watches the "me."
That is awareness.

This awareness is steady and non-judgmental. The "me" may struggle, succeed, overreact, or shine–but the "I" remains calm, clear, and centered. The "me" acts while the "I" observes–a quiet presence within us, like a third party who simply watches without taking sides. The "I" is our essence, the still point from which clarity and balance emerge. The soul sits in the I.

Living From the I

The "I" is more than an observer – it is the inner seat of centeredness, the place from which we lead ourselves.

From here:

- You can watch thoughts without becoming them.
- You can feel emotions without being consumed by them.
- You can witness situations without losing yourself in them.

Ordinary moments – a sunset, a melody, a walk – deepen when experienced from awareness.

Yet life constantly pulls us back into the me – the doer, the achiever, the fixer. And that's fine – until it drives us into automatic reactivity.

The key is learning to shift from "me "back to "I" – interrupting habitual patterns and choosing a wiser response.

A simple practice:

At the end of the day, pause and reflect:

Where did my "me" take over today?
What would my "I" have seen if it had paused?

That single question is how you begin leading from your center.

You Don't Swing at Every Pitch

When we watch a baseball game, we mostly notice the swing. What we rarely see is what happens before it–the quiet moment when the batter centers himself and decides whether to swing at all.

The ball comes fast. In a fraction of a second, the batter must decide. From the outside, it looks like a reflex. In reality, it isn't. Great hitters are *in the zone*. From that centered stance, they read the pitch and choose their response. Swing too early or too late, and you miss. Take the bait on the wrong pitch, and the result is an easy out.

Leadership works much the same way.

You wake up ready to start the day–and then everything unravels at once.
The internet is down. The water heater stopped working. The kids are late for school. Your phone keeps buzzing. And in a few hours, you're supposed to deliver an important presentation that still isn't quite ready.

If you start swinging at everything, reacting to each problem as it comes, chances are you'll exhaust yourself–and still miss what matters most.

Awareness changes that. It allows your "I" to take the helm before the "me" gets swept into unproductive activity. From that centered position, you can see the whole field, not just the next pitch coming at you.

Leadership isn't about swinging at everything life throws your way. It's about recovering your stance, staying grounded, and choosing which pitches are worth your energy. Only then can you act with clarity and increase your chances of hitting what truly matters.

Acting From Awareness: The Soul-Based Regenerative Leadership Framework

Surrendering and stepping into the vantage point of the observer enable awareness – and awareness is the prerequisite for leading from our best self.

We hinted at this in previous chapters, from the Leadership Diamond to Doing Less, Better. But going forward, awareness becomes even more central. It is the foundation of Soul-Based Regenerative Leadership – the red thread that connects each of the four steps.

Awareness is the entry point.
But awareness alone is not the destination.

Once we learn to observe ourselves with clarity and compassion, the next step is to channel that understanding into how we lead, decide, and create.

This is where Soul-Based Regenerative Leadership begins – transforming awareness into action, and action into renewal.

The Four Steps Ahead

1. **Skills – Doing with clarity**
2. **Mindset – Thinking differently**
3. **Soul Signature– Being from our essential qualities**
4. **Receiving – Leading from a higher source of insight**

These are not stages to check off, but movements in a continuous cycle – doing, thinking, being, and transcending.

The Soul-Based Regenerative Leadership Framework

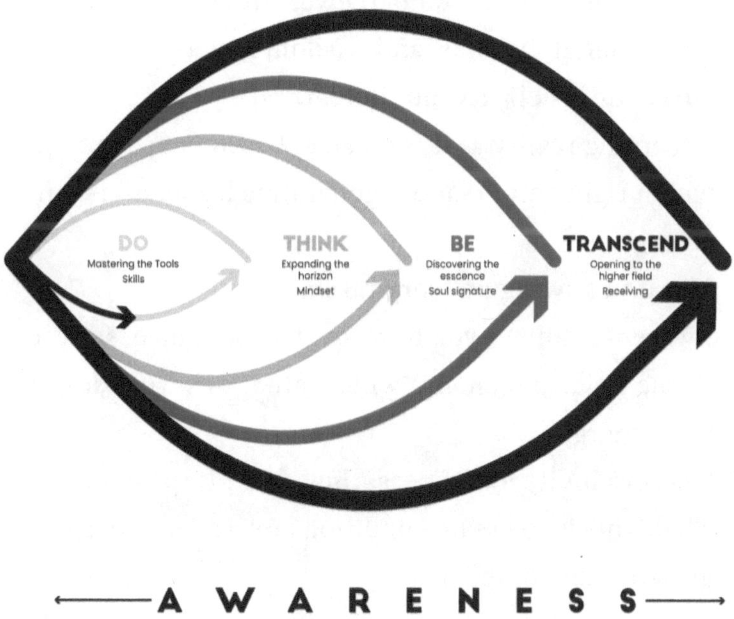

Reader Takeaways

1. **Awareness is your foundation.**
   It is the ground you stand on–the red thread that connects inner clarity with outer action and keeps you centered amid uncertainty.
2. **Awareness unfolds in two stages.**
   First, we let reality in without distortion. Then, we step into the vantage point of the observer and choose our response.
3. **You are both actor and observer.**
   The *me* performs; the *I* perceives. Integrating both expands discernment, balance, and wisdom in action.
4. **Surrender is clarity, not defeat.**
   Accepting reality as it is creates the inner space from which right action–and regenerative leadership–can emerge.
5. **You don't swing at every pitch.**
   Awareness allows you to recover your stance, see the whole field, and choose which situations truly deserve your energy.
6. **Leaders invite awareness; they don't impose it.**
   Their role is to create conditions for reflection, agency, and growth–not compliance.

7. **Awareness is the gateway to Soul-Based Regenerative Leadership.**
   It enables the unfolding of skills, mindset, soul signature, and receiving–the full arc of becoming your best self.

CHAPTER 6

Step 1 – Mastering the Tools: Skills

My grandmother was caught completely off guard when she heard the news.

"I'm sorry," the doctor said, "but we'll have to remove your daughter's eye."

She froze as the words sank in.

My mother had been born with a tiny black mark on the white of her left eye. It was small, but unusual. The doctor warned, "If you delay, her life could be at risk."

This was the 1940s, and the diagnosing physician was one of the most respected ophthalmologists in Mexico. Alarmed, my grandmother sought second and third opinions. Each one confirmed the same verdict: better to be safe than sorry. Remove the eye.

But she refused.

Something inside her–call it intuition, courage, or quiet conviction–told her to wait. Two years later, another doctor examined my mother and said calmly, "There's nothing to worry about. It's just a small pigment–like a mole. She'll be fine."

He was right. My mother kept her sight–and her life.

Moments like that remind us how fragile certainty can be–and how much depends on the quality of our knowing. But it's not only what we know that matters. It's what we *do* with that knowledge that makes the difference.

The gap between mastery and mediocrity can alter the course of a life, as my mother's story shows. In the AI era, tools will undoubtedly evolve. But the principle remains the same: the skills we bring–whatever form they take–still determine who advances and who falls behind; who stands out in an increasingly average world.

Having the right skills is the first step. It is the platform from which human capacity is released–and from which we rise toward our best selves.

Let's Talk Skills

The first step in the Soul-Based Regenerative Leadership framework begins with what you can do.

Your performance–how well you execute and how consistently you improve–is the foundation of your development journey.

In any activity you choose to pursue, you must be good at something–or at least good enough to meet a competitive threshold that allows you to stay in the game. Whether you're a teammate, a leader, or a customer, it's natural to expect a minimum level of acceptable performance. No one wants to deal with poor execution.

You don't start at your best–but you must start with enough capability to do a solid job.

If you enjoy music, you know exactly what I mean. Going to a karaoke bar can be great fun, and almost anyone can have a good time there. But chances are you wouldn't hire a karaoke singer to perform at your wedding. You can usually tell the difference between an amateur and a professional within the first minute.

Skills matter on the path to developing human capacity because they form the platform from which growth begins. Through them, you act your way into a new level of being. This is how growth becomes real–not in theory, but in practice.

Skills are the most visible layer of your contribution. They are the part of you the world can see, assess, and compare. Long before anyone asks about your mindset, values, or potential, your skills speak first.

That's why organizations pay such close attention to them.

Think about the last time you applied for a job, were considered for a promotion, or were evaluated for a new responsibility. Someone reviewed your profile. You were filtered–through algorithms, assessments, interviews, or all of the above. Weeks passed. You were tested, compared, and measured from multiple angles.

At its core, the organization was asking a simple question:

Do you have what it takes to deliver?

If you've been through that process, you know how exposed it feels–to be evaluated, ranked, and compared, hoping that what you've built so far is enough. You trust that your skills will carry you forward.

If things go your way, you move ahead. You're hired. But that's only the entry point. The clock starts ticking immediately. Now you must deliver, prove your worth, and justify the decision. If you don't, the system resets–and it's back to square one.

From the organization's perspective, this logic is unavoidable. Securing the right skills is a survival strategy. In highly competitive environments, companies go to extraordinary lengths to attract and retain capable people. The market for AI experts–with its staggering compensation packages–makes the point unmistakably clear:

Skill matters.

Yet beneath all the filtering, testing, and measuring lies a simple truth: being good at something–and being able to help others with it–is the minimum threshold of value.

That threshold is the ground on which everything else is built.

In the Soul-Based Regenerative Leadership journey, skills are the first step. They are not your identity, but they are your entry ticket–the way your inner capabilities become useful to others.

Work at them consistently, because lacking the necessary skill level will derail you.

The message is clear: you must strengthen your craft–because you are going to need it.

Skills as a Source of Joy

There is another side to skills–one that goes far beyond employability or performance. Skills can also be a profound source of joy.

When you're good at something, a positive reinforcing loop begins to form. Skill often generates happiness because you naturally enjoy the things you do well. And when you get to do what you love, something powerful happens: practice compounds–not only proficiency, but enjoyment as well.

The more you do it, the better you get. As you improve, you enter a virtuous cycle that fuels both mastery and fulfillment.

If you play golf, paint, run, or play an instrument, you know this feeling well. You practice, you notice progress, and that progress motivates you to keep going–to shave off a few seconds, hit a cleaner shot, or refine a detail just a bit more. Being good at something you enjoy opens the door to getting even better, which brings deeper satisfaction. Perfecting your craft may be part of your responsibility at work, but it can also become a springboard to happiness and self-realization.

This dynamic fascinated psychologist Mihaly Csikszentmihalyi. Through his research on the relationship between skills and challenges, he arrived at an extraordinary discovery: the state of *Flow*.

Flow is a state of complete absorption–the moment when you are so immersed in what you're doing that time seems to disappear and effort feels effortless. Action and awareness merge. You move with the task, and the task moves with you.

A musician lost in a melody.
A craftsman refining a detail until it feels just right.
An athlete so attuned that every movement feels instinctive.

What's remarkable is that Csikszentmihalyi wasn't studying performance when he described Flow–he was studying happiness. And what he found was simple and powerful: happiness emerges when challenge and skill are in harmony. Flow occurs when a task stretches your ability just beyond its current limit–about 10 to 20 percent more than what you can easily do.

- Too hard → frustration
- Too easy → boredom
- Just right → joy, creativity, and growth

This narrow band–the stretch zone–is where motivation, development, and satisfaction meet. It is also where people begin to feel most alive in their work. Getting better at what you do doesn't just elevate your performance; it elevates the quality of your life.

This is a solid–and powerful–first step in developing human capacity. The energy it generates doesn't remain contained; it naturally flows into the next stages of growth.

And while the journey to strengthen our skills begins within each of us, we still need an environment that nurtures that growth and supports our rise to our best selves.

This is where leaders can make all the difference.

How Leaders Can Help

In an era shaped by AI and constant disruption, leadership can no longer rely on an extractive performance model built on control, codified intelligence, and pressure. That model once delivered results–but it does so by consuming human capacity faster than it can be renewed.

A regenerative approach asks something fundamentally different of leaders. Not tighter control, but greater care for the conditions in which people contribute. Not pushing harder, but expanding the team's ability to learn, adapt, and grow by activating what is uniquely human. This is where leaders can make the greatest difference: by creating environments that build human capacity rather than exhaust it.

Part of the challenge is that the future of work is still being written. Many of the functional capabilities organizations will need tomorrow remain unknown–just as roles like data scientist or cloud architect once seemed unimaginable. New positions will continue to emerge as work evolves, and many will surprise us. Titles such as AI Workflow Curator, Hybrid Intelligence Facilitator, or Ethical Prompt Architect may sound speculative today, yet they point to a reality in which work is continuously being reconfigured.

While new technical skills will undoubtedly arise, distinctly human capabilities–judgment, empathy, imagination, adaptability, and presence–will only grow in value over time.

The heavy lifting of skill development ultimately rests with the individual. Commitment and effort are non-negotiable. But leaders play a decisive role in shaping whether that effort compounds or stalls. They do this not by directing every move, but by designing the conditions in which growth becomes possible.

In a regenerative model, leaders help by:

- **Supporting people through transitions**, recognizing that learning often requires letting go of familiar ways of working before new ones can take shape.
- **Co-creating evolving development paths**, not as fixed career plans, but as adaptive journeys that respond to changing demands, emerging skills, and individual potential.
- **Granting autonomy to apply peak skills while developing new ones**, enabling people to contribute from their strengths as they stretch into unfamiliar territory.
- **Acting as a sounding board**, offering perspective, questions, and guidance when required, without diluting individual responsibility.

- **Making progress visible**, acknowledging learning, experimentation, and skill acquisition–not just polished results–so growth is reinforced along the way.

The Quest Ahead

In helping our people rise into their best selves, it's worth acknowledging a simple truth:
skills matter.

Without them, we're not even eligible to begin. And as skills improve, the platform from which we grow becomes stronger and more resilient.

I once came across a remark attributed to Steve Jobs: a great software engineer can be orders of magnitude more productive than an average one–ten to fifty times more. The gap between good and great can be astonishing.

The encouraging part is this: you don't have to be a genius to close that gap. Small, consistent improvement compounds. Become just one percent better each month and the effect is dramatic–nearly one and a half times more effective in three years, and almost twice as effective in five.

This is how skills become an engine. As Winston Churchill put it, "*We shape our tools, and thereafter our tools shape us.*" Skills work the same way. We build them–and then they build us.

Staying relevant by renewing our skills is not a phase; it's a lifelong pursuit. And in a world that keeps rewriting the rules, the real challenge is not only to commit to that journey ourselves–but to help others rise along with us.

Reader Takeaways

1. **Skills are the entry point to regenerative performance.**
   Before mindset, soul signature, or receiving, you must be able to do something well enough to create value.
2. **There is a minimum threshold for relevance.**
   You don't need to be the best–but you must be good enough to stay in the game and committed to getting better.
3. **Skills turn effort into growth.**
   Through disciplined practice, performance improves–and with it, your capacity to become more than you were.
4. **Mastery can be a source of joy.**
   When skills meet the right level of challenge, flow emerges–where performance, fulfillment, and aliveness converge.
5. **Growth lives in the stretch zone.**
   Too much challenge creates frustration; too little leads to boredom. Sustainable growth happens just beyond comfort.

6. **Regenerative leaders build capacity, not pressure.** Their role is to create conditions where people can learn, experiment, and renew skills–rather than exhaust themselves proving value.
7. **Skill renewal is a lifelong pursuit.** In a world that keeps rewriting the rules, staying relevant is never finished. We must commit to our own development–and help others rise along the way.

CHAPTER 7

Step 2 – Expanding the Horizon: Mindset

I don't have what it takes to succeed. I'm going to quit. That was my conclusion after an honest assessment of the situation. I was alone in my office, and it felt unmistakably clear–I wasn't cut out for the job.

If you've ever questioned whether you were cut out for something, you've been here too.

It had been only three months since I was promoted to a senior-level position as Chief Human Resources Officer of the holding company that owned a business I had previously led. The organization employed more than 80,000 people at the time. Before that, I had successfully run a major business unit, so when they offered me this higher-level corporate role, I accepted without hesitation.

How hard could it be?

I was a novice in the role, and the challenge was massive: preparing the succession of all CEOs and most C-suite executives across the group, along with regional heads–more than 130 roles in total. But I didn't hesitate. I said yes.

In my defense, I had always been a fast learner. I believed hard work and heart could bridge any gap. And as a seasoned executive, I was used to solving complex problems.

Still, this time felt different.

After a few weeks of getting to know the field–attending seminars, meeting peers, and immersing myself in HR practices–I came to a painful realization: I wasn't "the HR type." I didn't feel I had the DNA for the role. The language, the mindset, the approach–all of it felt foreign. The job I thought I could master began to feel like a straitjacket.

Uncomfortable in my new skin, I seriously considered quitting. I even fooled myself into thinking I could return to my old role or find another position within the group. But deep down, I knew that was unlikely–and worse, it felt like betraying the trust my boss had placed in me.

I'm not a quitter. I've faced big challenges and overcome them. Yet here I was, stuck between a rock and a hard place.

Then, in the middle of that frustration, a thought surfaced that changed everything:

What if, instead of being the CHRO, I became the innovation guy in charge of HR?

The idea lit me up. It made perfect sense. I had always seen myself as an innovator–an outlier–someone comfortable exploring new paths. And the situation we faced in HR was unique. Traditional solutions weren't going to work. So why not bring that same spirit here?

That single shift in identity changed everything. It sparked a new belief I could stand on. Energy and joy returned. I felt confident, creative, and eager to act.

With this renewed perspective, I gathered my team, shared my epiphany, and invited them to see themselves differently too–not as a traditional HR function, but as an innovation team in charge of HR. Together, we could be game changers and reach ambitious goals.

It worked beautifully. We reimagined our function, designed fresh solutions, and delivered results that exceeded expectations.

That's the power of identity. When who you are aligns with what you do, energy flows. When you change the story about who you are, your horizon of possibilities expands–and you achieve more than you ever thought possible.

The experience has stayed with me ever since.

Embrace a new identity, and your mind–and the reach of your actions–will follow.

Let's Talk Mindset

Just as other people can't see what's happening inside your mind, you never really know what's unfolding beneath the surface of someone else's life.

While people are busy working on projects, nurturing relationships, caring for their families, cooking dinner, driving, shopping–simply moving through the world–something entirely different from what they appear to be doing is happening inside them.

They are living an inner life that is intense, complex, and full of twists and turns. And that invisible inner world is the real force behind their visible actions.

I would even argue that the time and energy we spend inside our own minds is often greater–and far more draining–than anything others see us doing. In any workplace, there are private dramas playing out every single day that no one else notices:

"Shoot... it's my turn to speak and I don't know what to say."
"There's no way I'm hitting this budget. I'm going to get fired."
"I can't stand this new colleague... and now we have to work together. Great."

And yet, on the outside, we appear to function and interact "normally."

In a very real sense, we all live a double life:

- the outer life everyone can observe, and

- the inner life that shapes everything we think, feel, and choose.

The quality of that inner life ultimately determines the quality of the outer one.
When we use our inner world well, our horizon expands–we grow, stretch, and rise.
When we leave it unexamined, it becomes one of the main reasons we stall.

And here's the good news: there is far more we can do to improve the quality of our thinking than we were ever taught.

So don't buy into the idea that mindset is some forbidden black box–something only therapists or specialists can open and fix. Yes, there are moments when expert help is needed. But for most of our mindset work, we are far more capable than we think.

All it takes is willingness, curiosity, and a simple method–as you'll soon see.

Look under the hood.
Get to work.

You are far more capable of "inner engineering" than you've been led to believe. Because the work we do inside ultimately determines how far we can go outside.

The Mind Is a Voice

If you want to improve how your mind works, you first have to recognize it for what it is:

an opinionated voice that never stops talking.

Can you hear it now?

There it is again–reminding you to go to the dry cleaners, pay the utilities, or buy groceries right in the middle of your reading.

And that voice doesn't work alone.

The mind has a full supporting cast at its disposal: images, memories, old conversations, things our parents told us, remarks from teachers who encouraged or dismissed us, comparisons with friends, moments of success we treasure, and failures we'd rather forget.

All of these fragments blend into ongoing narratives about who we believe we are.

And the inner voice finds ways to bring those narratives to life–often without our permission:

"Are you sure you want to be an artist? You've always been clumsy."
"Numbers? That's not your thing."
"Speak in public? Don't even try–you embarrassed yourself once in high school."

You know the drill. We've all lived versions of this soundtrack.

But the real question is: are these stories actually true?

They may be rooted in real moments, yes–but they only become permanent when we accept them as truth.

When I was growing up, I carried a story about myself that felt like destiny. My inner chatter pointed me toward working in government or maybe in the media. Those were the walls of my identity. That was who I believed I was.

I became the script.

There was no room in that story for business–much less for running one. And yet, a new story eventually took root. Here I was.

When a story repeats long enough, it stops feeling like a thought and starts feeling like who we are.
That's when a narrative quietly hardens into an identity.

So the real question becomes this:
If the voice in my mind shapes my identity, can I reshape that voice so it helps me expand who I am–and better seize the situations life presents?

To answer that, we have to look beneath the voice and its stories–to something deeper:
the realization that we can switch identities as we go.

You may think your identity is fixed–like a birthmark you carry for life.

But here's the truth:
it isn't.

We shift who we are depending on where we are and what we face.

Sometimes in the blink of an eye.

The Identities We Wear

Most of us grow up believing we have a single, fixed identity–a stable sense of self that shows up the same way in every situation. But in reality, identity is far more flexible and dynamic than we realize. We don't behave as the exact same person everywhere we go.

We speak differently at work than with friends.
We carry ourselves one way with our family and another with our peers or teams.
We naturally adjust how we think, act, and present ourselves depending on what the moment asks of us.

In truth, we each carry a repertoire of identities–different versions of ourselves that get activated by context and demand. Think about it:

- When you're having dinner with your in-laws, how do you show up?
- When you're presenting to your boss's boss, what version of you steps forward?
- When you're playing ball with your kids, what part of you comes alive?

These aren't fake versions of you.
They are all you–expressed differently.

Does that mean we're being inauthentic?
Not at all.

It simply means we have multiple identities to draw from, and–most of the time–we choose them intuitively. That isn't falseness. It's intelligence. It's adaptability. It's being human.

The key point is this: our minds and behaviors are far more malleable than we think.
We are not confined to a single way of being or trapped in one "default self."
We can choose the identity that best supports the moment–and still be fully ourselves.

Think about how we dress.
We wear a tuxedo to a wedding.
A T-shirt and flip-flops to the beach.
Different attire for different contexts.

Our identities can work the same way.

Imagine them hanging in an inner closet–ready to be chosen intentionally, depending on what the moment calls for.

We absolutely can do this.
But sometimes, stepping into a different identity requires fine-tuning the instructions running in the background.

Think about your phone. How often do you get notifications asking you to update your apps?

"Fixing a few bugs and improving performance."

Your mind works the same way.

To step more naturally into identities that expand your reach and performance, you may need to update your internal software–refresh old beliefs, install new ones, and clear out what no longer serves you.

When you do, new identities don't feel forced.
They feel available.
And sustainable.

That's why, to elevate performance and expand capacity, an Inner Reset becomes necessary–and that's exactly where we're headed next.

The Inner Reset: A Three-Step Method to Change Beliefs

We grow accustomed to being the way we are–even to the point of accepting traits we know don't serve us well. We adapt. We rationalize. We make peace with patterns that quietly limit us.

And then something happens.

A moment, a challenge, or a realization sparks a simple truth: if we want to do better, we can't keep running on the same inner wiring. Without change, we stay exactly where we are.

Every meaningful change needs a reason to begin. And the greater the change we're attempting, the stronger that reason–the *why*–needs to be. The good news is that we don't need to hit rock bottom to start. Wanting to improve–choosing to grow–is enough. Still, the first steps are rarely easy. Shifting our stories, identities, and mindset takes intention, and building momentum can feel surprisingly hard.

So where do we begin?

We begin with the basic building blocks of the mind:
our beliefs.

If you're looking for a domino effect–real, lasting change–this is where it starts.

But here's the catch.

Contrary to what most people think, we can't change our beliefs by confronting them head-on. Relying on willpower alone is a losing battle. It creates resistance, frustration, and very little real change.

You may already have a belief in mind–something you've struggled with, or something others have told you about yourself that feels like dead weight holding you back.
If so, hold it in mind as you keep reading.

If not, here's one that many people wrestle with–far more often than you might expect, even among young adults:

"I'm too old to reinvent myself. Old dogs don't learn new tricks."

Suppose you want to change it.

Your instinct might be to immediately replace it with a bold, determined declaration:

"Starting today, I'll learn everything I need to know. I can do this."

It sounds inspiring–and it may even give you a short burst of motivation.

But in practice, it almost never works.

Why?

Because deep-held beliefs don't collapse with a single punch. The mind doesn't shift just because we tell it to.

Strong beliefs–like this one–are anchored not only in logic, but in:

- emotion
- physiology
- identity

Trying to override a belief without addressing these deeper layers is like sweeping dust under a rug. It looks clean for a moment, but nothing has actually changed. The dust is still there–waiting. And sooner or later, it resurfaces, often at the worst possible time.

So if we want to change a belief–*truly* change it–we need a different approach. One that aligns with the natural pace and rhythm of the mind. One that works *with* the mind, rather than against it.

That's where the Inner Reset comes in.

It uses three simple but powerful steps–a counterintuitive sequence designed to unlock the mind's built-in ability to shift, expand, and rise.

Let's walk through them.

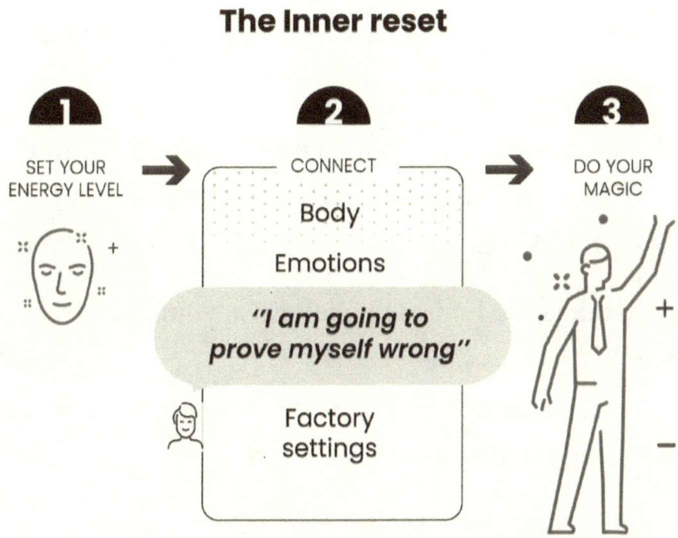

1) Set Your Energy Level

Before any belief can change, your energy must rise.

Have you ever tried to get something done–even something simple–when your energy is depleted? Suddenly everything feels heavier. Small tasks feel big. Your mind fogs up. You don't feel like yourself.

That's because you're operating in a suboptimal state.

Whenever you're about to put your mind to work, you want to start with extra battery, not a flashing red light. This doesn't mean becoming hyperactive or manic. It simply means elevating yourself to a state where you feel awake, present, and ready.

- Low energy keeps you grounded–in the worst way.
- High energy opens the door for change.

And the deeper the change you're seeking, the more important it is to sustain a steady, elevated level of energy throughout the process.

Think about sports teams before a crucial play. They don't stroll onto the field–they fire themselves up so they can step into the moment with full intensity and presence.

Before you take on a limiting belief, raise your energy a couple of notches.
It gives you the momentum you need to begin well–and the stamina to stay the course.

How is your energy level today?
Are you ready to work on your inner wiring–and make it better?

2) Connect

Body
It may sound counterintuitive, but the best way to access the mind is through the body first.

A composed body makes everything easier.
A tense body creates noise, constriction, and resistance.

And when you're not comfortable in your own skin, your capacity for new thinking shrinks dramatically.

This is why actors, athletes, speakers, and musicians all have rituals to loosen up before stepping into their craft. They know the body sets the stage for what the mind can do.

Limiting thoughts don't just stay in your head–they often "settle" in specific areas of the body. You can literally feel them as tightness, pressure, or blockage.

Let's return to our example:

"I'm too old to reinvent myself. Old dogs don't learn new tricks."

Chances are this belief–like many others–already lives somewhere in your body.

When you bring it to mind, notice where it lands:

- Does your stomach tighten?
- Does your chest feel heavy?
- Does your throat close?
- Do your shoulders tense up?

The mind can contract muscles and disrupt the natural flow of energy.

So the first step is simply to notice.

Ground yourself and:

- Take a few deep breaths–the proverbial ten deep breaths.
- Scan your body from head to toe.
- Locate the knots: the tension, pressure, or heaviness.
- Gently loosen that area.
- Breathe into it. Let the body soften.

This doesn't have to take long–just a few minutes of awareness can create a meaningful shift.

And yes, some areas may tighten again later. That's normal. When you notice it, release it again.

The goal isn't perfection.

The goal is reconnection.

Ask yourself:

Does my body feel ready to support me on the journey ahead?

Emotions

Beliefs and emotions are intertwined–you can't change one without affecting the other. And if you think you can reach emotions through logic alone, think again. Emotions don't yield to rational arguments. They require a different kind of approach because, truth be told, we're far less rational than we like to believe.

We are emotional beings who use rationality to justify what we already feel.

And this isn't just philosophy–there's solid science behind it.

Neuroscientists have discovered that when we make decisions, the emotional centers of the brain light up before the rational ones–like a Christmas tree turning on while logic hasn't even warmed up yet.

Emotions hit "go," and only afterward does rational thinking rush in to explain why we're already moving.

We love to believe we're thinking creatures who occasionally feel...
but the truth is, we're feeling creatures who occasionally think.

If we want to change a belief, we must learn to use emotions to our advantage.

We have two broad emotional buckets to draw from:

Empowering emotions
– curiosity
– enthusiasm
– courage
– determination
– confidence

Limiting emotions
– fear
– doubt
– shame
– insecurity
– overwhelm

Let's go back to our example:

"I'm too old to reinvent myself. Old dogs don't learn new tricks."

What emotion hides behind this belief?

Most people immediately say fear–and that's partly true. Venturing into the "new" exposes us; it feels risky.

But if you look more closely, you're not truly afraid of the unknown.
How could you be? You don't know it–it's hard to fear something you can't even define.

The real fear is the fear of leaving the known.

We cling to what's familiar because it feels safe–even when it quietly limits us.

This is why selecting the right emotion becomes essential.

Beliefs don't move on logic alone–they move when the emotion that fuels them changes.
To replace a limiting belief, you must also replace the emotion that keeps it alive.

For this belief, empowering emotions might include:

- excitement
- courage
- determination
- adventurousness
- curiosity

Any of these can give you the emotional lift needed to break out of the old story.

And yes–this can feel like a big leap. So here's a simple hack I use myself:

When I identify the belief I want to change–and the emotion underneath it–I smile, speak out loud, and say:

"This is what I think right now... but I'm going to prove myself wrong."

For example:

"I know I think I'm too old to change.
I know I sometimes resist new ways of doing things and fear I'll mess up.
I know I've convinced myself that old dogs can't learn new tricks...
but you know what?
I'm going to prove myself wrong."

And I don't say this with frustration or force. I don't say it with bitterness.

I say it with a smile–because a smile brings irony, lightness, and possibility.
It comes from a place of joy, not self-criticism.

So before you try to change a belief (and its twin emotion), pause and ask yourself:

- *What is the negative emotion fueling this belief?*
- *And what positive emotion can I choose instead?*

Then smile, remind yourself you're on your own side, and say with confidence:

"I'm going to prove myself wrong."

Factory Settings

Now we're ready for the real work of changing a belief.

It may feel daunting–some beliefs have lived inside us for years, even decades.

But here's the surprising part:

With the right approach, belief change can happen much faster than you think.

We've been taught that the mind is a complicated black box–a maze that's nearly impossible to rewire. But that's not the whole story.

Yes, the mind can be complex–
but the part that truly drives our behavior is far simpler than we tend to assume.

Most people operate with only a handful of "needle-moving" core beliefs–usually between four and eight. Just a few master rules run our internal operating system.

Which means this:

Shift even one or two core beliefs, and your entire inner world can change.
And when your inner world changes, your actions follow.

Let's revisit our example.

Restrictive belief:
"I'm too old to reinvent myself. Old dogs don't learn new tricks."

Now let's choose a new, empowering belief:

"The best is yet to come.
The greatest chapter of my life begins right now.
Every great deed begins small."

This new belief isn't just positive–it's foundational to rising to your best self.
It creates space, energy, and imagination.
It pulls you forward instead of holding you back.

To reinforce the shift, it helps to anchor it with a new identity–one that naturally supports the belief. Something like:

"I am a late-blooming, lifelong reinventor."

This becomes your new "default setting."

If you see yourself as a late bloomer who believes the best is yet to come, then changing at any age becomes not only possible, but expected. That identity opens doors that were previously closed. And even better, it strengthens who you are becoming–not just who you've been.

Put together, it sounds like this:

"I am a late-blooming, lifelong reinventor.
The best is yet to come.

The greatest chapter of my life begins right now.
Every great deed begins small."

Feel the difference?

This isn't wishful thinking–it's a new platform, a fresh launching pad for the life you want to create.

As you do this work, always adopt the stance of the "I" we explored earlier–the observing self, the chooser. From that vantage point, you can guide the "Me"–your operating self–toward the beliefs, emotions, and behaviors that move you forward.

Now that you've connected with your body, your emotions, and your new belief...

How ready do you feel to make it happen?

You've assembled a powerful combination.

Now it's time to bring it into action–
and begin harvesting the fruits of your inner work.

3) Do Your Magic

Now comes the moment to live from your new belief and identity.

You've raised your energy.
You've connected with your body and emotions.
You've reframed the belief.
You've anchored a new identity.

You're ready to act. And remember–*to think is to do.*

But life is life.

Not everything will flow smoothly the moment you begin using your new belief to guide your choices.
You will face headwinds.
You will stumble.
You will fail sometimes.

And that's okay.

What matters is how you respond.

This is where you "do your magic."

I borrowed this idea from something Carlos Alcaraz–the Spanish tennis star–once said in an interview. The interviewer asked him:

"Carlos, tennis is such a mental game. The hardest part is recovering after losing a point. What do you say to yourself when you're coming back from a mistake?"

Alcaraz smiled and answered:

"I tell myself, 'Carlitos... do your magic.'"

It's a wonderful way to talk yourself back into action–but what's even more remarkable is what he *doesn't* say.

He doesn't berate himself.
He doesn't say, "You idiot," or "What a terrible shot."
He doesn't drain his energy with shame or self-blame.

Instead, he speaks to himself with kindness.
With affection.
Using his nickname.

He lifts himself up.
He trusts that no matter how bad the moment looks, the magic is still there.

What a beautiful way to make a comeback.

So when you're working to change a belief–and life inevitably throws you a setback–be gentle with yourself.

Don't shrink.
Don't attack yourself.
Don't turn the stumble into a story of defeat.

Instead, pause.
Smile.
Say your name.
And finish with:

"...do your magic."

Then keep going.

Because you have it.
And you're proving it–one shift, one belief, one comeback at a time.

The Power of Faith

As we close the Inner Reset, there's one more element we need to talk about–one that often goes unmentioned but ultimately shapes everything that follows:

Faith.

Not in a religious sense (unless that speaks to you), but in the simplest, most practical sense:
the inner conviction that you can move forward even when you don't yet see the path.

As Dostoevsky once put it:
"Faith is not born from miracles, but miracles from faith."

In today's unpredictable world, logic and planning don't always carry us through. Sometimes what moves you from one step to the next is not certainty, but trust–trust in your abilities, trust in the process, trust in the possibility ahead.

Let's be clear about what this is *not*:

It's not blind optimism.
It's not crossing your fingers and hoping things magically fall into place.

It's a grounded kind of faith–the kind that says:

"I can handle this.
I've done hard things before.
I will figure it out."

Faith does three powerful things:

1. **It anchors your mindset when outcomes are uncertain.**
   When nothing is guaranteed, faith keeps you steady.
2. **It builds emotional resilience.**

It helps you stay centered during setbacks, mistakes, and slow progress.

3. **It sparks courageous action.**
When fear would normally hold you back, faith nudges you forward.

Faith cannot be faked.
It must be summoned–from experience, from courage, from hope, or from whatever source gives you strength.

And when your confidence wavers, faith steps in to bridge the gap.

So ask yourself:

Are you cultivating faith in yourself?
Do you have a daily practice–mental, emotional, or spiritual–that strengthens your belief that you can rise?

Because faith isn't optional on the road to becoming your best self, it's a daily choice.
And sooner or later, you will need it.

How Leaders Can Help

Mindset work is ultimately an inner, personal journey.
And yet–leaders play a crucial role in shaping whether people believe they can grow, or quietly give up on themselves.

This doesn't mean leaders should intrude on someone's inner process. But they *can* create the conditions, support, and presence that make growth feel possible.

A leader's words, gestures, feedback, and availability have the power to strengthen–or shatter–someone's inner narrative, often without realizing it.

Here are a few ways leaders can help:

1. Normalize inner struggles.
Everyone battles self-doubt.
And ironically, acknowledging this truth is often the doorway to becoming your best self.

When leaders speak openly about learning curves, reinvention, and growth pains, people feel permission to do the same.

2. Reinforce growth signals.
Pay attention to moments when people stretch beyond their comfort zone–even if the outcome isn't perfect yet.

Leaders don't just reward results; they highlight learning, adaptability, and persistence. Over time, this trains people to associate growth with progress–not punishment.

3. Help rewrite the inner narrative.
People are always telling themselves a story. Leaders influence that story whether they intend to or not.

Thoughtful language—especially in moments of doubt—can interrupt limiting beliefs and replace them with possibility. One sentence, delivered at the right time, can change someone's trajectory.

4. Walk the Inner Reset with your people.
Help them clarify what matters.
Encourage journaling, reflection, and conversation.
Share your own experiences of reinvention.

Reaching your best self—and helping others do the same—is a lifelong practice.
Make it a joyful one.

Reader Takeaways

1. **Identity shapes possibility.**
   Who you believe you are determines what you attempt—and what you avoid. Shift the identity, and new paths open.
2. **The mind is powerful, but not always helpful.**
   Your thoughts influence everything. Learn to guide them—or they will quietly guide you.
3. **Stories are not destiny.**
   The narratives you carry are interpretations, not truths. And interpretations can be rewritten.

4. **You don't have one identity–you have a repertoire.**
   Different moments call for different versions of you.
   Choose the one that serves the moment–and your growth.
5. **Belief change requires more than logic.**
   Energy, emotion, and body awareness open doors that willpower alone cannot.
6. **A single mindset shift can change your trajectory.**
   "I'm too old to reinvent myself" becomes:
   "I am a late-blooming, lifelong reinventor. The best is yet to come."
7. **Self-talk matters more than you think.**
   Treat yourself with the kindness of a champion.
   When you stumble, say your name and:
   "...do your magic."
8. **Faith is fuel.**
   You won't always see the path. But trusting that you'll figure it out keeps you moving forward.
9. **Inner resets drive outer results.**
   When you change how you think, feel, and see yourself, how you show up–at work and in life–changes too.
10. **You are far more capable of reinvention than you think.**
    Your best self isn't behind you.
    It's ahead–waiting.

CHAPTER 8

Step 3 – Discovering the Essence (Soul Signature)

Why is this happening to me?

That question kept echoing in my mind. I couldn't make sense of my situation. I felt helpless, confused, and completely overwhelmed.

It was midafternoon, and I was driving with no real destination–just looping through the streets, trying to think my way out of the mess I was in. My credit cards were maxed out. The restaurant project I had poured myself into had stalled because government approvals never came through. I had just become a father for the second time–this time to a child who needed medical treatment–and I had no job, no money, and no plan.

The outlook couldn't have been worse.

Traffic was heavy. I had already spent more than an hour going in circles with no clarity. Then, in the middle of my frustration, a simple thought crossed my mind:

I need a quiet place to think.

I remembered a small church nearby. *They're always open,* I told myself. *At least I can sit there in silence.*

I walked in, found a bench in the back, and sat alone. As in other defining moments of my life, I instinctively closed my eyes and turned inward–hoping there was an answer inside me that I had somehow missed.

The truth is, I felt completely lost. My self-worth was at zero. I couldn't understand why nothing was going my way. Friends around me were thriving, while I felt stuck at the bottom of a deep hole. I had hit rock bottom.

And I kept asking myself, again and again:

Why is this happening to me?

Then, as I sat there in stillness, something shifted. A sentence surfaced–clear, calm, unmistakable:

"You are supposed to bend, not break. The struggle is strengthening you."

It was an epiphany–an unexpected moment of clarity. Maybe my life wasn't meant to be easy. Maybe the difficulties I was facing were not punishments, but part of the process of becoming stronger so I could rise again.

Suddenly, my path felt less like failure and more like the forging of steel–intense heat and relentless hammering, not to break me, but to strengthen me.

With renewed resolve, I pulled myself together and began reviewing my options. I had to find a way–and eventually, I did.

What has always stood out to me about that experience is how, in our most crucial moments, we instinctively close our eyes to collect ourselves. We stop searching outward and begin searching inward.

We close our eyes when we need to think logically or solve problems. But when we face deeper questions–the ones about meaning, direction, and identity–we close our eyes, breathe deeply, and something subtle happens: the mind loosens its grip, and our attention settles elsewhere–often in the chest.

We go to our center.
We go to our soul.

At the most consequential crossroads of our lives, we turn inward and seek the seat of the soul. And when we reach it–even briefly–the noise subsides. The commotion stops. Things begin to make sense. The next steps become clear.

We reconnect with our core and move forward from that foundation.

The soul holds answers the mind cannot access. And our work, as leaders and as humans, is to learn how to reach that place, let it speak, and allow it to guide us–so we can live with meaning, clarity, and fullness.

Let's Talk About Soul

It's not intuitive to talk about the soul in daily life–much less in a business context, where the expectation is to operate through structure, skills, intelligence, and experience. These are practical, rational, and socially acceptable attributes. Our default setting is to lead with the "head." That's how we solve problems, seize opportunities, and make sense of the world.

But is that really the whole picture?

As valuable as logic is, it is only part of the story.

Nothing truly great has ever been accomplished without emotional commitment, inner conviction, and the fortitude that comes from being aligned within. When the compass inside us guides our actions, we tap into a deeper well–into the wisdom and energy that live at our human core. And when that alignment occurs, something shifts. A dimension opens that reason alone cannot reach.

Inhabiting the soul deliberately may not *look* practical, but don't be misled by appearances. The path to expanding human capacity–and meaning– requires it. You've likely experienced enough moments of inner clarity to know that something real, grounded, and powerful lives within you.

Once you recognize that this inner clarity is available to you, you begin to see that it's available to others as well. It's simply part of being human–and it shapes how we relate, connect, decide, and act together.

This is why the soul must become part of our daily conversation–and a guide for our actions.

Not as something mystical or abstract,
but as a natural, human source of direction.

And we must learn to ask ourselves–humbly and often:

What is my soul telling me to do?

And have the courage to trust that it knows the way.

The Subtle Path to the Soul

You may think your soul doesn't speak to you–that this territory feels too abstract, too "woo-woo." After all, you're dealing with real pressures: deadlines, budgets, expectations. Stepping into a "mystical" zone can feel out of place, even nonsensical.

And yet, I would bet you've had more moments than you realize when something inside you–call it your gut–quietly disagreed with your logical mind.

Think about the meaningful decisions in your life. How many times have you caught yourself thinking:

"This doesn't feel right. Something inside me is telling me to reconsider–or take a different route."

Maybe you left a well-paying job to pursue a dream.
Or chose a partner others didn't expect.
Or took a sabbatical instead of climbing higher up the corporate ladder.

Logic pointed one way–but something deeper pointed another. And you listened.

So what was speaking to you?

There may be psychological or philosophical explanations for that inner signal, but the lived experience is simple: something beyond your ordinary thought process was guiding you. In everyday

language, we already have a name for what happens when we look inward for direction:
soul-searching.

If it wasn't your mind–and it certainly wasn't logic–what else could it be? What other part of you can speak with that kind of quiet authority?

There is an extraordinary source of wisdom within us–subtle, precise, and often unmistakable once we learn to recognize it. But accessing it consistently requires more than instinct. It requires awareness, practice, and structure.

This is where **essential qualities** come in.

Your inner being is shaped by deep-rooted traits–qualities that, when expressed, make you come alive. Together, they create a distinct pattern of energy and expression. They reveal what is most natural and true in you.

That pattern is what I call your **soul signature**–the fingerprint of your soul, specific and uniquely yours.

When your actions align with your soul signature, life gains meaning. You feel on track. And just as importantly, you can sense when you're drifting–when you may be performing well on the outside, but moving away from yourself on the inside.

The path to surfacing your soul signature begins with an inward dive–starting with the discovery of your essential qualities.

Let's double-click on this next.

Essential Qualities: The Gateway to Your Soul's Signature

There is a distinct set of essential qualities within you–traits that give shape to what your soul longs for. They form the imprint of your essence: the signature of who you are at your core. They're already there, waiting to be surfaced and strengthened–so they can enrich your inner life and elevate how you perform.

If you're unsure what yours are, here's a simple exercise to bring them into focus.

Think back over the last couple of weeks. Recall the tasks and moments you lived through–projects, assignments, presentations, daily chores, hobbies, volunteer work, or meaningful conversations. There's no need to go further back; recent memories are clearer.

List a few moments that stand out. For example:

- Solving a problem for a customer
- Helping a colleague improve at work
- Painting a picture
- Negotiating a deal
- Creating a time-saving process for your team
- Doing volunteer work
- Delivering an impactful presentation about a product launch
- Walking in the park with your spouse while making plans for the future

Now choose **three to five** moments from your list.

Next, identify which of those were genuinely enjoyable or fulfilling–moments that brought a spark of happiness, a sense of rightness, or the feeling that what you did mattered. Select the ones you'd gladly repeat.

Let's say you chose these three:

- Helping a colleague improve at work
- Delivering a presentation about a product launch
- Walking in the park with your spouse while making plans for the future

Now distill what made each one meaningful. Ask yourself:

- What made this feel good?
- What part of me felt engaged?
- What exactly was meaningful about it?

Write down the elements that energized you. For instance:

In helping your colleague:
bringing clarity, creating order amid complexity, serving as a sounding board.

In delivering the presentation:
distilling complex ideas into clear insights, expressing yourself, connecting with your message.

In the walk with your spouse:
reconnecting emotionally, envisioning a shared future, nurturing your relationship.

Now look for patterns–as if you were reverse-engineering your fulfillment. Ask:

- What qualities were present in those moments?
- Which parts of these experiences do I consistently enjoy?
- What feels uniquely "me" when I'm at my best?

You may notice themes such as **clarity, curiosity, communication, gratitude, service, creativity, empathy, wisdom, courage**, or **harmony**.

These are clues.
Don't just rationalize them–*sense* them.

Highlight the qualities that appear again and again, and choose **five to eight** that feel most essential.

These are your essential qualities.

They signal who you are and what your soul resonates with. And they are one of the most direct pathways to connecting with your soul.

Take the time to do this exercise on paper. Let your reflections settle, then return to them. Sit with your essential qualities. Notice where they live in you. Notice how naturally they arise–and how it feels when you express them.

Your essential qualities form the bundle that expresses your soul's signature. As you use them consistently, you grow with them. Over time, they strengthen and become a launching pad–the seat of your soul.

When you live from these qualities, you're not just doing work you enjoy. You're aligning with your inner self. You're allowing the deepest part of you to express itself fully.

The soul leaves clues.
Your job is to follow the breadcrumbs.

Your essential qualities become your inner compass.
They echo what your soul wants for you–
and guide you toward your best self.

The Soul's Canvas

In the previous exercise, we began with tangible actions–real tasks, real behaviors–and traced them back to what lies beneath: the qualities that give your work meaning.
Once you've identified your essential qualities, the next step is to nurture them.

At first, this process may feel subtle. You may not know exactly how to express your essential qualities in daily life–and that's natural. Most of us have been operating from a dominant program for years, one built around habits, expectations, and external demands. It takes time to reset.

As you begin noticing your essential qualities–and experimenting with them–small sparks of curiosity or inner energy start to appear. Stay with those sparks. Like an acquired taste, they grow slowly at first... and then, almost without warning, they begin to infuse many of the things you do. Over time, they become your natural way of performing.

Think of the soul as a canvas, and your essential qualities as brushstrokes. In this metaphor, you're not trying to replicate someone else's painting–you're revealing a landscape in your own style.

Each time you notice a feature of that landscape and act from one of your essential qualities, you add a brushstroke. And with each stroke, something becomes clearer:

- what feels right,
- what genuinely expresses you,
- what wants to take on a life of its own.

As the painting develops, the strokes begin to connect. They build on one another. Gradually, a coherent picture emerges–almost as if this is what the canvas had been waiting for all along.

Imagine your essential qualities include gratitude, communication, service, creativity, or empathy. These become the brushstrokes that bring your canvas to life. When you use them deliberately in your daily work, they begin to guide your actions intuitively.

- If you're preparing a report, express service. Look for the real need behind the request. Seek feedback. Go the extra mile–not for perfection, but for relevance and meaning.
- If you're delivering a sales pitch, express empathy. Connect before you convince. Let the human relationship open the door to the business conversation.
- If you're onboarding a new team member, express communication. Create a two-way exchange. Ensure clarity, and make space for the other person to feel heard and seen.

These moments are brushstrokes.
They define your style.
They express who you are.

As you continue painting with your essential qualities, you remain open to new ideas–but you also begin to recognize what resonates as distinctly yours:

- your style,
- your approach,
- your way of engaging,
- your recurring themes.

Over time, these elements become unmistakable.
They give your work–and your life–an identity.
They reveal your essence.

And this is when something important happens:
your soul is no longer something you search for.
It becomes something you express.

Start painting. Let your essential qualities surface, and allow the soul to guide your hand.
The canvas has a way of revealing where it wants to go.

The Passion Paradox

What about passion?
Where does it fit alongside essential qualities?

Isn't passion supposed to be the key to discovering who we are meant to become–our true path? We're told that if we follow our passion, everything will fall into place. *Find what you love, and you'll never work a day in your life.*

Well... yes and no.

In real life, passion–especially in a work context–is far more elusive than the slogans suggest. Outside of work, in hobbies or personal pursuits like art, volunteering, sports, cooking, or reading, passion is easier to recognize. People lose themselves in these activities because something inside them comes alive.

But at work?

Not so much.

Time and again, I've seen how difficult it is for people to experience what they would genuinely call passion in their professional lives. Engagement data reflects this reality: most workers report being disengaged in some form.

Part of the problem lies in how we approach the search for passion.

We tend to treat passion as if it were a high-performance vehicle–one capable of accelerating from zero to one hundred miles per hour in seconds–expecting that once we find it, we can simply step on the gas and let it carry us forward.

If you've found such a vehicle, great. They do exist–but they are rare.

And when we build our expectations around finding that kind of engine, frustration is almost inevitable. Most people spend years searching for the role, industry, or calling that will propel them to top speed, only to discover that it either doesn't exist–or doesn't behave the way they hoped once they find it.

And this is where the paradox becomes clear:

The harder you try to find the one job that ignites your passion, the more it tends to slip away.

Passion is not a fixed destination.
It is a moving signal.

It shifts with age, with responsibility, and with the season of life you're in.
What energizes you at twenty-five may not energize you at forty-five.
What feels meaningful today may feel irrelevant tomorrow.

If you're one of the fortunate few who has found a passion and built a life around it–wonderful. Truly. Keep going.

But for most people, chasing passion becomes frustrating, exhausting, and deeply confusing. Passion moves. It evolves. Sometimes, it even burns itself out.

Here is what I've learned:

Working with your essential qualities–surfacing your soul's signature–is far more accessible, more stable, and far more fulfilling than chasing passion.

Returning to the painting metaphor: instead of forcing yourself to decide what kind of painter you *should* be, let the brushstrokes–the essential qualities–reveal who you already are. Build from there.

Passion may come and go.
But essential qualities stay with you.
They grow with you.
They guide you.

And every time you act from them, your soul expresses its signature.
That is what it means to lead from the seat of your soul.

How Leaders Can Help

This is where leadership transcends skills and roles.

Your task is no longer just to direct performance, but to help others discover, express, and strengthen their essential qualities. When leaders operate from this place, they unlock possibilities no process, incentive, or system ever could.

Imagine the compounded effect inside an organization where people can identify their soul's signature and consistently perform from their essential qualities.

What would work feel like if individuals led from the seat of their soul–liberating human capacity–instead of operating from pressure, fear, or compliance?

As a leader, your responsibility is twofold:
to bring your best self to work by surfacing your essential qualities, **and** to create the conditions for others to do the same.

Leading from your core becomes your mandate.

1. The Seat of the Soul – Locate It Within You

Ground yourself in your soul's signature.
Lead from your essential qualities–not your title.

When you recognize your own soul, you begin to recognize the soul in others.
This is the moment leadership shifts–from transactional to transformational.

2. Speak to the Soul, Not the Role

Your team members are not their job descriptions; they are human beings on their own journey of becoming.

As Napoleon observed:
"It is by speaking to the soul that man is electrified."

People give their best not for money or perks, but because something inside them has been awakened.

3. Amplify from the Inside Out

Help people express their essential qualities.
Give them space to use their strengths.

If someone must design processes but thrives in creativity, let them innovate.

If someone wants to develop empathy, place them in human-centered situations.

When people act from their essential qualities, they become energized, committed, and deeply engaged.

4. Create Opportunities for Soul-Searching

Make yourself available to help people explore their core–if they choose to.

Encourage reflection and self-awareness.
Be a facilitator rather than an evaluator.

Help people discover who they truly are–not just who the org chart says they should be.

5. Connect as an Ally, Not Just a Boss

Show your team that you are with them–not to shield them from challenge, but to help them grow through it.

Challenges become developmental opportunities when people feel supported, seen, and understood.

Soul-Based Regenerative Performance

When leaders operate from the seat of their soul, they unlock the soul of the organization.

People rise not through pressure, but through alignment.
They perform not from obligation, but from conviction.

And when essential qualities are nurtured collectively, the entire system begins to regenerate–from the inside out.

As technology takes us into uncharted territory and the boundaries of human contribution continue to evolve, the mandate becomes unmistakably clear:

Lead from your soul.
Let your essential qualities guide you.
And help your teams navigate their own soul-searching journeys.

If you take nothing else from this chapter:
notice which qualities give you energy–and choose to lead from them deliberately this week.

Reader Takeaways

The soul is your deepest compass.
Logic solves problems, but the soul provides meaning, direction, and identity–especially at life's most consequential crossroads.

1. **Your soul signature is the core of who you are.**
   It is the unique pattern formed by your essential qualities–the traits that make you come alive and signal alignment or drift.
2. **Essential qualities are the gateway to the soul.**
   By identifying and expressing them, you gain reliable access to inner clarity and a stable foundation for action.

3. **Passion is an unreliable guide.**
   It shifts with time and circumstance. Essential qualities remain steady and can bring meaning to any role or season of life.
4. **Soul-connected living is a practice.**
   Alignment strengthens through repeated recognition, expression, and use of your essential qualities in daily work and life.
5. **Your soul's canvas is revealed through action.**
   Each expression of an essential quality is a brushstroke that shapes who you become and clarifies your path.
6. **Soul-based leadership elevates performance.**
   When leaders amplify essential qualities and speak to the soul, engagement and effectiveness rise from the inside out.
7. **Organizational regeneration begins within.**
   When individuals act from essence rather than pressure, the system becomes more human, resilient, and alive.
8. **The path inward is the path forward.**
   The more aligned you are with your soul signature, the clearer your decisions become–and the closer you move toward your best self.

CHAPTER 9

Step 4 – Opening to the Higher Field

"I trust your negotiation abilities."

It was Saturday, 5:00 a.m., and I was sitting at my kitchen table after a restless night. In a few hours, I was scheduled to meet our second-largest client at their headquarters–and they were furious.

That sentence kept echoing in my head.

A week earlier, our best-selling product had been delisted from their chain. It was retaliation for a scheme my boss had put in place to curb excessive discounting. If the retailer offered deep discounts, their access to other products would be restricted–products their competitors could still carry. On paper, the intent was clear: protect the perceived value of our leading products and help everyone make more money in the long run.

In reality, it did not sit well with them.

Angry doesn't begin to describe their mood.

They threatened to kick us out altogether. They said we were no longer a reliable business partner. The relationship was at a breaking point. Their legal counsel even hinted at a lawsuit, claiming the scheme was illegal. We weren't just losing the account–we were exposed.

They rejected my repeated requests to meet until they finally agreed–on one condition. I was to show up alone. My boss was explicitly not welcome. On their side, they would bring the purchasing manager I worked with, his supervisor, and the Vice President of Sales–the second most powerful executive in the organization, just below the CEO.

It was David versus Goliath.

My boss was unequivocal. He would not back down. My task was to convince the client to accept the scheme and reinstate the business. The only thing he offered me was a sentence:

"I trust your negotiation abilities."

"But how am I supposed to fix this?" I replied. "We created the problem, and now the client is furious. We can't keep pushing your scheme–it's leading us straight into a major confrontation."

He smiled and repeated, "I trust your negotiation abilities."

This wasn't reassurance.
It was a warning.
Get it done–or deal with the consequences.

I had no strategy. Every argument I rehearsed in the days leading up to the meeting ended in a dead end.

That morning, sitting at the kitchen table, I made myself a coffee. As I took the first sip, I stopped thinking. The warmth, the aroma, the simple act of being there pulled me fully into the moment. I found myself in stillness.

And then–out of nowhere–an idea appeared.

I opened my laptop and reviewed the client's historical pricing data for the product at the center of the conflict. The numbers were clear. There had been a couple of close calls, but according to my boss's own scheme, they had never violated the lower threshold. They were fully compliant.

For all practical purposes, the "problem" had never existed.

When I arrived at the client's meeting room, they were already there. No greetings. Only the Sales VP spoke.

"This will be a short meeting," he said. "We're not complying with your crazy plan. We can find other suppliers. Thank you for your business." He began to stand.

"I'm sorry this escalated," I said. "That was never our intention. But we're divided over a problem that never actually occurred."

"What do you mean?" he snapped. "We received your notification in writing."

"Yes, you did," I replied. "But check your pricing history. Whether the plan exists or not is irrelevant–you're well within the thresholds. You've never been in default. It's business as usual. It doesn't make sense to fight over a non-existent problem."

The VP turned to his team. "Is that true?"

They paused. "We'll need to double-check."

He took out his phone and called one of their analysts. After a few minutes, the answer came back.

"It sounds right."

He looked at me and smiled. "See? I told you it was going to be a short meeting."

He straightened up. "If there's no problem, let's forget about it and get back to work."

That was it.

Things returned to normal.

I don't know where the idea to frame the situation as a fight over a non-existent problem–and to back it with just enough evidence–came from. It didn't feel like thinking or devising a solution–it felt like receiving. The moment I stopped forcing and opened myself to the situation, the insight emerged quietly, like a whisper.

Was I the one who thought this?

And more importantly–how could I do this more often?

Let's Talk Opening to the Higher Field

Before you dismiss the idea of opening to the Higher Field as something that doesn't apply to you–or feels off-limits–I want to draw your attention to a familiar, almost mundane ritual. One that, in my experience, people reference more than any other when talking about sudden clarity or unexpected insight.

The shower.

If I had a nickel for every time a leader told me that their best ideas arrive while taking a shower, I'd probably have enough to fund several start-ups. It's happened to me often enough that I simply smile when I hear others describe the same experience. Not long ago, while talking with a friend about these so-called "shower moments," he told me he has them all the time–so often, in fact, that he bought a waterproof notepad and pen designed specifically for that purpose, just to capture ideas before they vanished. He used to keep it just outside the shower. Still, sometimes it was too late. By the time he stepped out and dried off, the idea was already gone.

The point is simple–and I suspect deeply familiar.

Most of us have lived moments like this. You're carrying an unresolved problem, something you've been turning over in your mind without making progress. And then, without warning, a solution appears. Fully formed. Uninvited. As if it came out of nowhere.

It doesn't feel like thinking.

In fact, the thinking mind seems momentarily suspended–softened by the ritual, distracted, almost absent.

What it feels like instead is *receiving*.

This phenomenon isn't limited to everyday life; it has shown up repeatedly throughout history, across disciplines and centuries. Take Archimedes, for example. He is said to have discovered the principle of buoyancy while stepping into a bath–another shower moment. As he noticed the water rise, something clicked. He suddenly understood how to measure the volume of irregular objects. Legend has it that he was so exhilarated by the insight that he jumped out of the bath and ran through the streets of Syracuse shouting, "Eureka!"–"I've got it!"

The French mathematician Henri Poincaré described a similar experience. While boarding a bus, he later wrote, *"At the moment when I put my foot on the step, the idea came to me, without anything in my former thoughts seeming to have paved the way for it."* That spark eventually led to his work on Fuchsian functions.

And of course, there's Isaac Newton, who reportedly arrived at the foundational insight behind gravity while quietly contemplating in a garden and watching an apple fall from a tree.

Artists tell similar stories. Musicians like Mozart spoke of hearing entire compositions in their minds–complete and fully formed–long before writing them down. Their work, in a sense, was simply to transcribe what had already arrived. And the list goes on.

The pattern is remarkably consistent. Insight does not always arrive through effort or linear reasoning. Sometimes it appears when the mind loosens its hold–when we stop forcing answers and become available to receive them. It's as if information reaches us from beyond the boundaries of our usual thinking, from a field that sits just outside conscious control.

That is what I mean by *opening to the Higher Field.*

What matters for leaders is not simply acknowledging that these moments exist–we've all experienced them–but recognizing that they are not random. There is an underlying logic to them. A pattern. And if we understand that pattern, we can learn to access clarity more intentionally.

To liberate human capacity, we can connect to a higher field of consciousness–one from which clarity and wisdom naturally emerge.

So let's slow this down and look more closely at what that actually means.

Receiving from the Higher Field

Look around you for a moment–and pause.
We are so accustomed to our surroundings that we often lose sight of something that is, quite literally, mind-blowing.

The room you're in. The car you drive. The street you walk on. The supermarket where you shop. Everything you see shares something remarkable in common. Life is everywhere–animated, organized, and coherent. Whether visible or microscopic, simple or complex, living systems surround us at all times. They are structured, ordered, and guided by an inner intelligence.

Life expresses itself in countless ways. Yet beneath that diversity lies a consistent pattern: every living thing follows a path true to its nature. A flower moves through the seasons expressing its flowerness. A bird migrates and lives by expressing its birdness. Each organism fulfills its own calling, and at the same time, all are interconnected–networked into a larger whole that functions together.

There is an underlying system that organizes life–a master architecture that gives it direction and coherence. This is what I call the Higher Field: an intelligent arrangement that allows everything to function independently while remaining deeply interconnected.

A helpful way to think about it is as an operating system. You don't see the OS running on your phone or computer, yet it quietly coordinates how all the apps and programs work together. In much the same way, the Higher Field is the "software" of life–the invisible structure and energy that binds everything and keeps it in motion.

And in any complex system, coordination requires communication.

Feedback loops, signals, and continuous adjustments are what allow a system to stay alive, adaptive, and coherent. Modern physics offers a quiet reminder here: reality does not operate only through what is immediately observable. Invisible fields and relationships precede form. What once seemed implausible is easier to accept now: much of reality is shaped by relationships we don't directly see.

The Higher Field works in the same way. Its intelligence is not static; it is dynamic and responsive. What we experience as insight, intuition, or inspiration often goes beyond pattern recognition or the free association of ideas alone. It arises when we tune in to a deeper organizing system already at work within us.

This Higher Field is always on.
It transmits continuously, like a signal that never shuts off. The wisdom is there, available to be received–but only if we are tuned in. And here's the key: connecting to this signal doesn't require sophisticated "Bluetooth" or "Wi-Fi." It requires attention.

But there's a catch.

The signal does not arrive loudly.
It whispers.

Its guidance is subtle–easy to miss and quickly drowned out by noise: pressure, urgency, overthinking, constant stimulation. That's why insight so often arrives when the mind relaxes–while showering, walking, driving, or sitting quietly. When the noise drops, reception becomes possible.

So how do we improve our "signal strength"?

We create an interface with the Higher Field.

That interface develops when we reconnect with what the soul longs for–when we return to our own nature. The antenna is formed by our essential qualities, the unique signature that allows us to resonate with a wider source of wisdom. When we operate from those qualities, our bandwidth expands. We align with who we truly are and who we are meant to become. In that alignment, coherence with the larger system becomes possible–and insight flows with less effort and less friction.

Ideas don't feel forced.
Clarity doesn't feel manufactured.
Decisions don't feel heavy.
They feel received.

This is why living and leading from your essential qualities matters so deeply. As we discussed in an earlier chapter, when you express them consistently, you are not only acting in ways that feel meaningful–you are tuning your inner instrument. You become more receptive to coherence, guidance, and clarity from the Higher Field.

You can think of this as a quiet dialogue–your soul attuning to life itself.

In everyday language, we call this inspiration. Sometimes we say we were "visited by the muses." Another word we use is intuition. We let go of control, and suddenly an idea appears. It feels like it came out of nowhere. We catch it, test it, shape it, and build on it.

The question is how to invite that kind of reception more often.

Connecting to the Higher Field - The Practice

Up to this point, we've explored what the Higher Field is and how it shows up in moments of insight, clarity, and unexpected breakthroughs. The natural question that follows is a practical one: **How do we access it more intentionally?**

Opening to the Higher Field is not a one-time event or a mysterious gift reserved for a few. It is a **practice**–one that can be cultivated. While each person may experience it differently, the process tends to follow a recognizable pattern. At its core, the connection unfolds through a series of stages that move us from inner readiness to outward action.

Opening

Opening begins with recognizing our essential qualities–the signature of our soul–and allowing them to resonate fully within us. We open by suspending judgment, quieting internal commentary, and letting go of the emotional and cognitive baggage that narrows perception.

This is an inward turn. A deliberate pause. A shift in attention.

By choosing to be present rather than reactive, we adopt a receiving posture–clear, grounded, and attentive. This step is foundational. When opening is done well, everything that follows becomes easier and more natural.

Allowing

Allowing is the opposite of forcing.

Rather than pushing for answers or trying to control outcomes, allowing means creating interior space–making ourselves a vessel. It is the natural continuation of openness, where we let what wants to emerge do so in its own timing.

This requires setting ego aside–not to diminish ourselves, but to loosen our attachment to control and certainty. From this place, our essential qualities begin to lead. We resonate less with who we think we should be and more with who we are meant to become.

Doing – Downloading

Insight does not remain abstract for long. It seeks expression.

There is a quote attributed to E. M. Forster that captures this stage well: *"I do not know what I think until I hear what I say."* As we articulate–whether through speaking, writing, sketching, or playing–ideas begin to take shape and often surprise us.

You've likely experienced this. As you talk through an idea, words emerge that catch you off guard: *Did I really just say that?* In those moments, thinking is no longer separate from doing.

Creators understand this intuitively. Writers, painters, and musicians often describe being guided by the work as it unfolds. Doing is not just execution; it is another way of receiving. Action becomes a dialogue with the Higher Field, where clarity continues to emerge through movement–not before it.

The next time you need to design a plan, write an email, or prepare for a meeting with your boss, resist the urge to overanalyze before acting. **Think enough to establish direction, then begin.** Start with a good-enough draft and let the work itself tell you where it needs to go. As you move, ideas emerge, connect, and find their way forward.

Action doesn't replace thinking–it **enriches** it.

Consistency

As with any meaningful practice, repetition matters.

The more you return to these stages–opening, allowing, downloading, and doing–the more natural the process becomes. Your bandwidth widens. Your trust in the practice grows. Over time, connecting with the Higher Field feels less like an exception and more like a reliable source of guidance.

At that point, the question shifts from *whether* insight will come to *how often* you choose to make space for it.

And If You Listen Very Hard, the Tune Will Come to You at Last

Once you begin to notice how this process works in your own life, you start to recognize it elsewhere–sometimes in unexpected places. One of the most powerful expressions of this shows up in a song many of us already know.

The words in this heading come from one of the greatest rock songs ever written: *Stairway to Heaven*–a classic that has resonated across generations. Written by Robert Plant and Jimmy Page of Led Zeppelin, it is a song that has remained iconic for more than five decades. I'm far from a music expert, and I can't fully explain why *Stairway to Heaven* has endured the way it has–but I suspect it has something to do with the inner journey the song invites us to take, through both its lyrics and its music.

Art often has a way of getting there first–sensing and expressing truths long before we find the language to describe them. *Stairway to Heaven* does exactly that. It can be interpreted in many ways, revealing different meanings depending on where each of us is in our own journey. What caught my attention in the context of this book is how, in its own way, the song captures the essence of opening to the Higher Field and the kind of wisdom that emerges from it.

At its core, the song invites us to question what we usually chase. It opens by challenging the assumption that material success–"all that glitters is gold"–should be the ultimate goal, and gently redirects our attention toward a deeper dimension of ourselves.

Along the way, it reminds us that things are not always what they seem. A sign on the wall may look clear, yet words can carry more than one meaning. The invitation is to go beyond appearances, to look past surface interpretations, and to search for what is true rather than what is merely obvious.

As the song unfolds, it becomes clear that the "stairway to heaven" is not something external to be climbed, but an inner journey–one that begins in the soul and opens into understanding. *"And if you listen very hard, the tune will come to you at last."* Clarity, the song suggests, does not emerge from forcing or striving, but from listening–tuning in to a deeper intelligence, the kind of wisdom the Higher Field represents.

Even the music mirrors this movement. It begins quietly, almost like a whisper, and gradually builds in strength and intensity. It feels as though something deeper is guiding the ascent–subtle at first, unmistakable later. And when all the elements come together, when "all are one and one is all," we arrive at a state of alignment: grounded, present, and connected to the larger consciousness of which we are a part.

That, to me, is the quiet power of the song. It takes us on a journey of opening and receiving–of listening deeply, moving inward, and reconnecting with a source of clarity, meaning, and direction. In that sense, the stairway is not something we climb, but something we open.

In the end, what the song points to is the importance of creating enough inner space for what is already within us to come forward–liberating human capacity rather than constraining it. From that place–seated in our soul's signature–we open ourselves to a wider field of intelligence that guides growth, coherence, and becoming. When we listen very hard, we don't just hear the tune–we find our way forward, rising to our best self at the human frontier. And in that sense, "buying a stairway to heaven" is not about possession or achievement, but about choice–the choice to create space, to listen, and to commit to the inner journey.

How Leaders Can Help

Opening to the Higher Field is ultimately a personal journey. It cannot be mandated, taught mechanically, or forced. That said, leaders play a crucial supporting role. When teams are willing, leaders can help create the conditions that make receptivity, insight, and clarity more likely.

The key is presence without intrusion – being available, not directive.

Here are several ways leaders can help:

1. Normalize "letting the genie out of the bottle"

Not all insight comes from linear analysis or preplanned logic.

Leaders can normalize this by openly acknowledging that ideas often emerge unexpectedly – while walking, reflecting, or in the middle of conversation. Naming this reality gives people permission to trust those moments rather than dismiss them as unprofessional or unreliable.

When leaders legitimize this way of thinking, fear and self-censorship decrease. Insight feels welcome – even when it arrives imperfectly formed.

Practical cue:
Ask questions like, "What's emerging, even if it's not fully clear yet?"

2. Encourage practices of reflection and self-awareness

Reflection is not a luxury; it is a leadership capability.

Leaders can encourage simple practices such as journaling, quiet reflection, or brief pauses after intense work. These moments help reduce internal noise and strengthen self-awareness – the foundation for receiving deeper insight.

Reflection also creates space for ideas to surface, evolve, and take on a life of their own.

Practical cue:
Build short pauses into meetings or transitions and signal that reflection is part of the work, not a break from it. Ask, "What insight are you taking away – what needs time to sink in?"

3. Create safe spaces for expression and action

Insight needs expression to take shape.

Leaders can encourage people to speak up, experiment, and move into action without over-policing ideas too early. When teams feel safe articulating half-formed thoughts, unexpected connections emerge.

This doesn't mean abandoning rigor – it means sequencing it. First expression, then refinement.

Practical cue:
Separate idea generation from evaluation. Make it explicit when a conversation is about exploration versus decision-making.

4. Recognize the process, not just the outcome

Opening to the Higher Field is not about being right; it's about becoming receptive so that, over time, we get it right.

Leaders can reinforce this by recognizing curiosity, effort, and engagement with the process – not just successful outcomes. There is no single "correct" way to access insight. What matters is familiarity and trust in the practice.

Over time, this builds confidence and reduces fear of failure.

Practical cue:
Acknowledge thoughtful attempts and learning moments, even when results are still unfolding. Say things like, "We're onto something – let's let it grow and see where it leads."

5. Model the behavior

Perhaps the most powerful signal a leader can send is personal example.

When leaders speak openly about how they pause, reflect, listen, or struggle to find clarity – and how insight eventually emerges – they make the journey visible and human. Transparency turns abstraction into permission.

People don't follow instructions; they follow signals.

Practical cue:
Share short, real stories of how clarity emerged for you – not just the final decision, but the path that led there.

6. Protect space from constant urgency

The Higher Field rarely speaks in moments of constant pressure.

As much as is feasible in the hectic environments we live in, leaders can help by protecting small pockets of space – time without immediate demands, meetings without predefined answers, and conversations without a rush to closure.

This is not about slowing down performance. It is about improving the quality of decisions.

Slow down to speed up.

Practical cue:
Design at least one recurring space where the goal is sense-making, not execution.

Closing

In this way, leaders don't open the Higher Field for others. They help remove what stands in the way. They create the conditions where insight can surface, coherence can emerge, and people can rise – individually and together – liberating human capacity in the process.

Reader Takeaways

1. **Some of your best insights don't come from thinking harder, but from letting go.**
   Clarity often emerges when pressure drops and the mind relaxes–when we stop forcing answers and become available to receive them.
2. **Opening to the Higher Field is not random; it follows a pattern.**
   Insight tends to arise when we move through a recognizable practice: opening, allowing, expressing through action, and returning to it consistently.
3. **Your essential qualities are not just who you are–they are how you receive guidance.**
   When you lead from your essential qualities, you are not only acting authentically; you are tuning yourself to receive clearer, lighter, and more aligned insight.
4. **Insight whispers–it doesn't shout.**
   The Higher Field communicates subtly. Noise, urgency, and overthinking drown it out. Stillness, reflection, and presence make reception possible.
5. **Doing is part of receiving.**
   Insight rarely arrives fully formed. As you speak, write, build, or act, clarity continues to emerge. Action is not the end of the process–it is part of the dialogue.

6. **Rising to your best self is an inner journey before it is an outer one.**
   Like the Stairway to Heaven, growth is not about climbing something external, but about opening inward–creating space for what is already within you to emerge.
7. **Leadership is not about having all the answers, but about making space for them to emerge.**
   When leaders learn to open–to themselves and to others–they create the conditions where coherence, insight, and wiser decisions surface naturally

CHAPTER 10

Claiming the Soul's Edge: A Soul-Based Regenerative Leadership Path

"There is nothing more powerful than
an idea whose time has come."
–Victor Hugo

A few years ago, while working on a company-wide transformation initiative, my then-boss–the CEO–asked me to gather ideas on what should come next. Many of the initiatives already in motion had delivered real value to the business, yet subtle signs of strain were beginning to surface. Some newer efforts were colliding with established policies. People expressed uncertainty about what was expected of them–and about how much freedom they truly had to act within existing boundaries. Something needed to change, and he wanted direction.

In early August, he asked me to imagine the next chapter of the transformation and gave me a couple of months to think it through.

"Come back with something that stretches our thinking," he said.

We scheduled a meeting for October to review my recommendations. But for various reasons, the meeting was postponed–again and again.

I immersed myself in the work. I spoke with key stakeholders to understand what was troubling them, distilled the patterns I was seeing, and pushed my thinking beyond incremental fixes. I did exactly what I had been asked to do: stretch. Still, it wasn't until January of the following year–nearly six months after the initial request–that we finally met.

After the presentation, I summarized the direction I believed we should take.

The CEO looked at me and said,

"It's clear you've worked hard on this. You've given it a lot of thought."

"Yes," I replied. "I'm genuinely excited about this work. It captures the few critical moves we need to make to regain our step. Some ideas still need refinement, of course, but we have time to polish them and define a timetable for deployment."

He paused, then said,

"Thank you for the presentation. For now, put it in a drawer. There's a lot to think about. We'll revisit this later."

And that was the end of it.

My work was never revisited. I was never told why. Perhaps the proposal was considered too bold, or simply failed to resonate. Postponement felt safer. In hindsight, the repeated rescheduling of the meeting should have told me enough.

A year or so later, the organization was forced to respond. The situation demanded a course correction and a different approach. By then, however, I was nearing retirement. The moment had passed. I never learned what might have happened had the proposal been given the green light. Another day at the office.

Whether my proposal would have worked, I will never know. But I do know this: kicking the can carries an opportunity cost. Inaction imposes a hidden burden–one that is rarely visible at first, but becomes undeniable once problems grow large enough to demand attention. I've since learned that moments like this are not failures. They are invitations–to ask what evidence the other person needs in order to change their mind.

I share this story because, more often than not, timing is everything. Even strong ideas can be set aside indefinitely–not because they lack merit, but because the moment is judged inconvenient or uncomfortable.

Perhaps you've experienced something similar. You see the road ahead with clarity while others hesitate. You wonder how something that feels obvious to you is not obvious to them. That frustration is familiar–and deeply human.

But take heart: truly game-changing ideas are rarely comfortable. If an idea feels entirely plausible from the outset, chances are it isn't bold enough. When you bring forward something genuinely different, some people will think it's crazy–and that is often a sign you are onto something. Familiar territory rarely produces real innovation or lasting change.

This is where the conversation about the soul enters.

If you feel hesitant to speak about the soul–your own or that of your people–in a business environment, you are not alone. Yet the conditions are unmistakably pointing in that direction. The old extractive performance model has reached its limits. To meet the challenges of the AI era, we need a different logic–one that restores human capacity rather than depleting it.

That is why the time for Soul-Based Regenerative Leadership has come.

The invitation is clear: we must act sooner rather than later. Opportunities do not wait indefinitely. Catching the wave at the moment it can carry you forward makes all the difference.

Victor Hugo was right. Learning to recognize when an idea's time has come may be one of the most important decisions a leader can make.

How the Framework Works in Practice

The four steps of the Soul-Based Regenerative Leadership framework can be understood in two complementary ways.

The first is developmental and sequential: a progression in which each step builds upon the previous one, creating momentum and guiding growth over time.

The second is integral and regenerative: a perspective in which the sequence reveals something deeper–that the soul is the gravitational center of the journey.

Awareness anchors both views. It is what enables depth, coherence, and meaningful transitions between stages.

The developmental lens: a sequence that creates momentum

At its core, the journey follows a simple yet powerful progression:

Do → Think → Be → Transcend

This sequence follows a clear inner logic. Skills help us reach a baseline level of performance. Without a solid craft–rooted in doing–outstanding results are simply not possible.

Yet skills alone are not sufficient. A second step is required: mindset. Distinctive skills must be supported by beliefs and attitudes that fuel adaptation, learning, growth, and resilience. Strong capabilities cannot flourish if they are not supported by a mindset that invites challenge and advancement. Step one is amplified by step two. Together, they form a foundation for performance–but the journey does not end there.

Performance expands further when it aligns with our essential qualities – the core of who we are becoming. At this point, the soul signature becomes fundamental. By reconnecting with these qualities, we begin to clarify who we are meant to be. This connection brings coherence and meaning to what we do and how we think, elevating our actions beyond execution toward sustained, superior performance. The soul becomes the seat from which alignment with our essence takes shape–unlocking levels of energy and contribution we may not have thought possible.

In the final step, we are invited to open ourselves to receive from a higher field–one that nourishes our essence and expands our perspective. From this connection, creativity and insight allow us to act with greater clarity and contribute in more distinctive ways. This is the territory of game-changing ideas and one-of-a-kind creations.

The Soul-Based Regenerative Leadership Framework

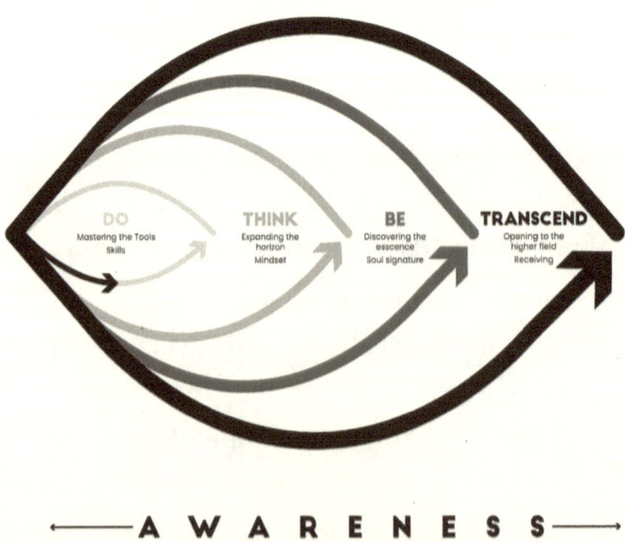

The integral lens: the soul as the gravitational center

There is, however, another way to understand the framework. When viewed integrally, the sequence reveals a deeper truth. What appears to be a linear progression is, at its core, the recognition that the soul is the gravitational center of the journey.

Doing and thinking are not the destination; they are preparation. They shape the vessel that allows the soul to express itself fully. Skills and mindset prepare the ground, giving form and structure to what follows. When they are aligned with our essential qualities–our soul signature–they create a solid foundation from which we operate.

From this center, we are able to open ourselves to receive from the higher field we have been describing. As the soul resonates, coherence deepens, wisdom becomes accessible, and action gains clarity, meaning, and direction.

The sequence helps us travel the road of development, but the true power of the framework lies in this recognition: it is the soul-based perspective that brings the model fully to life.

This dynamic extends beyond our personal journey toward expanding human capacity. It becomes exponential when we recognize that others are on the same path. Our aspiration is to elevate performance–our own and that of those we lead–through a regenerative approach. In a world that is constantly changing and demanding growth, how we undertake this challenge matters.

That is the central insight of the framework: the soul is the hidden engine of human capacity–and human capacity is the true value-creating force.

Pause for a moment and consider this:

What changes when you recognize the people you lead as souls on a journey?
What becomes possible when the conversation moves from role to role–to soul to soul?

Why This Approach Is Regenerative

What makes this framework regenerative is not the language, but the effect it produces. Unlike extractive performance models that consume human capacity to generate results, this approach restores and expands capacity through use. As skills deepen, mindset evolves, alignment strengthens, and connection to higher wisdom grows, energy is renewed rather than depleted.

There is no endpoint to this process. Increased proficiency invites deeper exploration, which strengthens the whole system and fuels further growth. Over time, development becomes a self-reinforcing cycle: growth generates capacity, and capacity enables further growth.

This is what regeneration means in leadership – performance that strengthens the human system that makes performance possible.

Do's and Don'ts

Although the Soul-Based Regenerative Leadership framework is a deeply personal journey, a few do's and don'ts can help you stay grounded as you engage with it.

Do's

Acknowledge that everyone starts where they are.
There is no right or wrong entry point. Each person begins from their current level of awareness, proficiency, and readiness. What matters most is the willingness to begin.

Honor the sequence.

Skills, mindset, soul signature, and connection to the higher field build upon one another. Skipping steps–or reversing the order–weakens the process. Depth and coherence emerge when the progression is respected.

Choose progress over perfection.

There is no such thing as perfection in this work. Growth happens through consistent, meaningful steps–not through idealized standards or unrealistic expectations.

Allow revisiting and deepening the process.

The journey is not linear in lived experience. Returning to earlier steps is not regression; it is part of regeneration. Each return deepens understanding and strengthens the whole.

Don'ts

Avoid the temptation to measure everything.

Conventional methods designed to quantify performance can be counterproductive here. Progress through the four steps cannot be fully captured by scores or checklists. As Jorge Luis Borges suggested, this work lives at the intersection of "adventure and order." Treat it as such.

Don't go halfway.

This journey requires full presence and commitment. Partial engagement drains energy rather than restoring it. Becoming your best self demands dedication; nothing of lasting significance has ever been achieved without it.

When the Soul Enters the Boardroom

Even when senior leaders are committed to a regenerative approach to performance, the shift will not take hold unless it is also embraced at the board level. Boards, like the organizations they steward, are exposed to the profound changes AI is bringing–and they cannot remain on the sidelines.

The board exists to steward the organization's future: to guide direction, oversee leadership and risk, and safeguard long-term value creation. It carries a fiduciary responsibility to shareholders–and increasingly to the broader ecosystem in which the organization operates. At its core, the board's mandate is to ensure the organization's health, continuity, and capacity to endure.

Yet this era poses an unavoidable question: **how does the board's role evolve in the age of AI?**

Traditionally, board members have been selected for their experience, judgment, and functional expertise–qualities assumed to be scarce and therefore valuable. Today, AI can augment analysis, scenario modeling, pattern recognition, and insight at unprecedented speed and scale. Much of what once defined expertise is becoming increasingly accessible.

So what, then, is left for the board to do?

What AI cannot do is carry accountability, steward meaning, or hold the human system together through complexity and uncertainty. This is where the board's distinctive contribution must evolve.

In an era defined by uncertainty, acceleration, and profound change, the board's role must expand–from overseeing performance to actively stewarding the conditions that make performance regenerative. Guiding people toward their best performance is no longer a peripheral concern; it becomes a strategic imperative. This is where soul-based conversations enter the boardroom.

And they do not apply only to the people within the organization.

When the soul enters the boardroom, the conversation extends beyond employees' journeys through the framework. It also includes the inner life of those around the table. Directors and executives alike are invited to reflect on how they themselves are showing up–and on what it means, in this context, to *liberate human capacity* and steward it wisely.

As AI increasingly commoditizes expertise and analytical insight, the amplification of human contribution moves to the forefront. Beyond reviewing the P&L, cash flow, or balance sheet, the board's role must include helping the organization regenerate the very capacities that make sustained performance possible. That requires a fundamental shift in perspective: recognizing the human soul as the true center of gravity for value creation.

In this new landscape, boards that wish to remain relevant–and competitive–must cultivate a Soul-Based edge. Not as a substitute for sound strategy or disciplined execution, but as an economic force multiplier. By leading from awareness–and fluently stewarding–the **Do → Think → Be → Transcend** journey, both within the organization and within themselves, boards position their companies to navigate complexity and uncertainty in the AI era.

Reader Takeaways

1. **Timing is a leadership capability.**
   Insight alone is not enough. Even strong ideas can fail if conditions are not ready. Leadership requires sensing when an idea's time has come–and having the courage to act when it does.
2. **Transformation begins with how leaders see people.**
   Before asking how to change, leaders must examine their assumptions. Moving from roles to souls fundamentally reshapes how performance, growth, and contribution are understood.
3. **The framework works in two dimensions at once.**
   It is both sequential–guiding development through **Do → Think → Be → Transcend**–and integral, revealing the soul as the gravitational center that gives coherence to skills, mindset, and action.

4. **Skills and mindset are necessary–but not sufficient.**
   They create a platform for performance. Fulfillment, coherence, and distinctive contribution emerge only when leaders reconnect with their soul signature–who they are becoming.
5. **Regenerative leadership restores energy instead of consuming it.**
   Unlike extractive performance models, this approach creates a reinforcing loop: growth expands capacity, capacity fuels further growth, and energy is renewed rather than depleted.
6. **Human development has no finish line.**
   None of the steps has a fixed endpoint. Revisiting earlier stages is not regression–it is regeneration. Progress invites deeper exploration, not completion.
7. **The ultimate competitive edge in the AI era is human becoming.**
   As AI commoditizes expertise and analysis, advantage shifts to those who steward the human system. Leaders and boards who guide the Do → Think → Be → Transcend journey–within the organization and within themselves–are best positioned to navigate complexity and uncertainty.

PART III

Crossing the Human Frontier

After simplifying the leadership landscape in Part I and exploring the regenerative performance approach in Part II, we arrive at the threshold of the human frontier. The next step is not conceptual–it is experiential. It requires crossing.

In this final part of the book, we examine what the future demands from leaders, organizations, and the system as a whole–and why the response must be fundamentally human. As technology accelerates and artificial intelligence reshapes how work is done, leadership can no longer rely on the assumptions that once guided performance and value creation. The direction of evolution is not toward greater control or intelligence, but toward a deeper human orientation – one grounded in coherence rather than control.

The chapters that follow explore three essential crossings of the AI era.

The first is the leader's crossing–the shift from hero to steward.
A personal transition that asks leaders to release the identity built on certainty, answers, and control, and to embrace a role centered on cultivating awareness, connection, and collective intelligence. Leadership becomes less about carrying the burden of knowing and more about creating the conditions for others to rise.

The second is the organizational crossing–the shift from coupling people with AI to leading beyond intelligence itself.
An organizational transition that reframes work and performance around what remains distinctly human as intelligence becomes abundant. Organizations must move from extracting results through pressure and control to generating them through human capacity, alignment, and meaning.

The third is the systemic crossing–the emergence of a new performance logic.
This crossing challenges the invisible equation that has governed organizations for decades. As codified intelligence becomes widely accessible, the old levers of performance lose their effectiveness. A different logic must take shape–one anchored in the expansion of human capacity and the integration of the human dimension into value creation itself.

The first crossing transforms the leader.
The second transforms the organization.

The third transforms the logic by which performance itself is produced.

In the AI era, the human frontier is not defined by more information or faster analysis. It is marked by a transition from knowledge workers to sentient workers–people whose contribution comes from awareness, judgment in uncertainty, and the ability to act from their best selves.

This transition–from extractive performance to regenerative performance–will not occur through a clean break. It will unfold while organizations continue operating within legacy systems that, if left unchanged, ultimately undermine the very stability they were designed to provide.

Leaders therefore face a genuine dilemma. Waiting for a fully proven method to navigate this transition may feel logical, but it is no longer feasible. In a world moving too fast for certainty to arrive first, leadership must adopt a different stance–one guided not by guarantees, but by discernment and conviction.

A different question must lead the way:

Is this true enough to bet on?
The AI era will demand that leaders discern, moment by moment, where to stand between non-human-centered work and human-centered work–and place human capacity at the forefront of value creation.

This is the crossing that lies ahead–the Soul Edge.

CHAPTER 11

The Leader's Crossing: From Hero to Steward

"Me neither."

A colleague shared a story not too long ago that has stayed with me. It captures the leadership moment we are living in.

She had reconnected with a friend of hers, a seasoned regional CEO–the kind of leader who had seen it all. Decades of success. A reputation for getting results under pressure. Someone who had navigated tough times before.

They hadn't spoken in a couple of years, and she was genuinely looking forward to the conversation. She cared about him and wanted to catch up.

She asked, "It's great to see you. How are you doing?"

He paused. Took a breath. Then said:

"I'm not sure what to say. The business isn't doing well. The market keeps throwing one challenge after another at us, and we're barely keeping up. We're doing our best, but in all honesty... for the first time in my life, I don't know what to do."

This was not what she expected to hear.

He went on to explain that after hitting a wall, he had requested a meeting with his boss–a global executive, widely admired for his experience. He thought he might find some guidance there.

In the meeting, he laid everything out. The challenges. The decisions he had made. The strategies he had tried. And the reality that none of it was working.

He ended by saying:

"I've tried everything. And I have to admit that, for the first time in my life, I don't know what to do."

His boss listened. Paused. And then replied:

"You know what? Me neither."

Imagine the setting.

Two accomplished leaders. Decades of experience between them. The kind of leaders who were expected to have it figured out. And suddenly–

No clear map.
No proven solutions.
No idea what to do next.

This is the world we live in now.

Intelligence, experience, and effort were once enough to get results. They were the silver bullet that helped us find our way out–but not anymore.

When I heard this story, I realized the leadership we've known is past its expiration date.

We are living in a moment where the old recipes have reached their limit. A different way forward is required. We need a new approach.

This is where **Soul-Based Regenerative Leadership** begins.

Heroless Leading

When everything is going well, leaders often look like heroes.

Results are strong. Execution is smooth. Targets are met. And success has a way of masking underlying defects. Teams appear to function like well-oiled machines, delivering consistently and efficiently.

But are they?

On the surface, leaders can usually explain why things are working. They can point to the strategy that outsmarted competitors, the innovation that unlocked growth, the cost reductions that protected margins, or the decisions that captured market share. These stories are compelling. They are also familiar. And they quietly reinforce what I call the **heroic halo effect**–the belief that performance flows primarily from the leader's brilliance, drive, and answers.

The problem is not that these stories are false.

The problem is that they are incomplete.

When results turn adverse–and sooner or later they do, now faster and more frequently than ever–the limits of heroic leadership begin to show. The very capabilities that once delivered success–decisiveness, speed, personal insight, relentless execution–start to lose their effectiveness.

The first cracks rarely appear in the numbers.

They appear in interactions.

Conversations change. They narrow. They become more reactive and less generative. The questions being asked–and the ones no longer asked–signal what is missing. Collective intelligence, once assumed to be present, no longer seems to make a difference. What follows, if left unaddressed, is often a gradual but unmistakable nosedive.

When teams underperform, their ways of working are usually diminished as well. Over time, people grow accustomed to being guided by the leader's epiphanies–by their clarity, their answers, their way forward. The unintended consequence is subtle but significant: the team's **human-capacity muscle** weakens. Initiative declines. Perspective narrows. Meaning thins out.

To avoid this condition, leadership must move beyond a purely heroic stance and be balanced with stewardship. The task is no longer just to deliver results, but to cultivate human capacity and support renewal.

One way to surface what is happening beneath the surface is to look through the lens of the four steps of the Soul-Based Regenerative Leadership framework:

- **What is the level of awareness in the room?**
  To what extent are people operating on autopilot, reacting out of habit rather than presence?
- **What skills are being applied and developed–and why?**
  Is there clarity about the real problem the team is trying to solve and the tools required to address it, or is the team defaulting to activity without understanding?
- **What mindset is shaping decisions?**
  Are people willing to look under the hood and question assumptions or beliefs that may no longer be serving them? What are we not thinking that could matter now?
- **How connected are team members to their essential qualities–to their soul signature?**
  Are they amplifying what they are naturally meant to be, or performing roles that drain energy and dilute impact?
- **And finally, how are they showing up in the world?**
  Is there room for inspiration, creativity, and intuition to inform problem-solving? Are people open to insight that emerges when logic and experience alone fall short?

Under the old leadership paradigm, the leader carried the burden of having the answers–the visionary, the hero, the knight on a white horse expected to ride in and save the day.

But in this moment–and even more so in the future–something else is required.

There will always be times when leaders must be decisive and take charge. That remains part of the role. But it can no longer be the default state.

The default is **stewardship**: creating the conditions for awareness, connection, and alignment so others can rise to their best selves and collective intelligence can emerge.

In doing so, the leader does not disappear. On the contrary, the leader's role becomes more demanding–and more consequential. By shifting from hero to steward, leaders help the team surface its best and, in the process, bring forth the best in themselves as well.

This is the new leadership role.

Not to extract value from people by pushing harder because you have the answers–but to regenerate value by cultivating human capacity, so the system itself becomes stronger, more human-bound, and more alive over time.

Being Comfortable in Your New Skin

The shift implies that you will have to live with a certain kind of grief.
Not grief as drama – grief as release.

The quiet ache of letting go of what once worked: the habits that brought results, the reflex to tighten control, the comfort of being the one with answers, the identity built around competence under pressure.

Old habits don't disappear with a single break, as we have seen earlier. Versions of you will linger longer than you expect. They will resurface in moments of stress or uncertainty. When they do, notice them. Hold steady.

This is not a cosmetic transition or a leadership "upgrade." It is a deeper reconfiguration. In the AI era, the cost of staying in the old equation is not just the risk of occasional underperformance. It goes further than that. It is the steady depletion of human capacity – the very resource your organization will depend on most.

So don't aim for perfection. Aim for steadiness.

This is not a leap you take once; it is a crossing from hero to steward you make again and again, until the new way of leading begins to feel like your own skin.

Because the role itself has fundamentally changed.

The hero carries performance.
The steward cultivates the human capacity that makes performance possible.

This is the crossing.

One practical way to support it is to treat it as a question-led inner process – something you return to deliberately, especially when you feel yourself slipping back into old defaults.

Use these questions as a compass:

- **What is holding me back from advancing?**

Identify the two or three obstacles creating the most friction right now – outdated assumptions, unresolved tensions, misaligned priorities, or structures that no longer serve their purpose.

- **What is trying to emerge and wants to be nurtured?** Notice the signals pointing to what matters most at this stage – new possibilities or directions that feel both necessary and timely, even if they are not yet fully formed.
- **What is ready to rise that I can now step into?** Clarify the role or stance you are being called to assume – and the few aligned actions that would allow you to embody it and move your journey forward.

These questions are not meant to be answered quickly. They are meant to be revisited.

Because the point is not to add more effort – it is to remove friction. And as you do, progress begins to unfold with greater clarity, coherence, and a deeper sense of alignment with who you are becoming.

Crossing Thresholds

The Soul-Based Regenerative Leadership framework is meant to serve as a canvas–one that helps leaders orient themselves and understand how the journey unfolds. It is not designed to depict growth as a sequence of disconnected steps. While each step requires focus and depth, leadership is lived as a continuous journey, one in which transitions matter as much as the steps themselves.

Progress does not come from mastering individual elements in isolation, but from integrating them over time. Awareness is what allows leaders to recognize these transitions–to sense where they are coming from, where they are heading, and when a shift is required.

The journey is much like a long hike. Imagine you and your team venturing into unfamiliar terrain together. You may be well prepared, yet you still need to adapt to conditions you could not anticipate. Different sections of the path call for different paces. Some stretches require sustained effort; others demand patience and restraint. There are turns, changes in elevation, and moments when the summit briefly comes into view. Each segment matters, but the journey only makes sense when seen as a whole.

As a leader, your responsibility is not only to reach the summit, but to help the entire team get there–and return safely. Leadership, in this sense, is a collective endeavor. There is no isolated win. Nobody wins until everybody wins.

As leaders cross thresholds, the center of attention shifts. The journey often begins with what needs to be done–developing skills and capabilities. Over time, attention moves to how situations are interpreted, and how mindset and beliefs shape decisions. From there, the process turns inward, resonating with the leader's unique qualities and how they can elevate performance. Eventually, awareness expands beyond the self, into a deeper sense of connection and wisdom.

These are not leaps.
They are thresholds–crossed as awareness evolves.

Seen this way, the steps of Soul-Based Regenerative Leadership are distinct, yet carried by a single stream. Leaders do not leave one step behind when moving to the next; each becomes integrated into how they lead. Awareness holds the journey together, guiding the movement from doing, to being, to connecting.

The Soul-Based Regenerative Leader is called to create a different kind of flow than the one we commonly refer to. Not the skill-based flow we are used to–where mastery, repetition, and execution take center stage–but a **Soul-Based Flow**.

In this flow, essential qualities become the gravitational center. Skills and mindset are still present, but they integrate into a deeper orientation–one in which the soul sets the direction. Performance, then, is no longer about force or effort alone, but about how the leader's best self–and the best selves of their people–are expressed through everyday actions.

Carrying the Journey Forward

Before we leave this chapter, I invite you to pause and reflect on your leadership journey so far. Whether you have spent years in leadership roles or are in the process of stepping into one, you have already covered meaningful territory–territory worth revisiting. It has likely included both accomplishments and shortcomings, moments of pride alongside moments of doubt. You have navigated difficult terrain, learned along the way, and here you are, still standing, ready to face what comes next.

One thing you can be certain of is that unforeseen challenges will inevitably arise.

The future will have plenty in store for all of us. The question now is not so much what you will do to move forward, but how. How much of yourself are you willing to bring into the journey ahead? And how much of others are you prepared to carry along?

You will not have all the answers. No one does. But becoming a Soul-Based Regenerative Leader helps you ask better questions–of yourself and of others. Questions that create clarity, invite wisdom, and open space for others to rise with you.

You are human among humans, a soul surrounded by souls. Let that certainty guide your way forward.

Reader Takeaways

1. **Leadership today requires a different source of strength.** When experienced leaders find themselves without answers, it signals not a lack of competence but a shift in context that demands a new way of leading.

2. **The defining leadership crossing is from hero to steward.** Sustainable performance now depends less on having answers and more on creating the conditions for awareness, connection, and collective intelligence to emerge.
3. **This transition requires releasing an old leadership identity.**
   Letting go of control reflexes, answer-driven authority, and the comfort of being the expert allows a more regenerative form of leadership to take root.
4. **Leadership growth unfolds through thresholds, not steps.** Progress comes from integrating skills, mindset, and inner alignment as awareness evolves and leaders cross into new ways of seeing and acting.
5. **Regenerative progress happens through subtraction.** By removing friction, outdated assumptions, and misalignment, leaders create the conditions for renewal and sustained forward movement.
6. **Soul-Based Flow aligns performance with human becoming.**
   When essential qualities become the gravitational center, leadership shifts from force to presence – allowing the best of the leader and the best of their people to emerge in everyday action.

CHAPTER 12

The Organizational Crossing: Leading Beyond Intelligence

"Prove It"

In the movie *Contact* (1997), there is a quiet–but devastatingly powerful–moment that captures a tension we rarely name, yet live with every day – the tension between evidence and what cannot be measured.

Dr. Ellie Arroway (played by Jodie Foster) is a brilliant scientist who has devoted her life to the search for extraterrestrial intelligence. She is rigorous, uncompromising, and deeply committed to empirical proof. If something cannot be measured or verified, it does not belong in the realm of truth.

Palmer Joss (played by Matthew McConaughey), on the other hand, is a man of faith. Thoughtful, grounded, and unafraid of ambiguity, he represents a way of knowing that does not rely on instruments, equations, or data.

At one point, as Ellie is being questioned about her beliefs–and her refusal to accept anything that cannot be proven–Palmer gently shifts the conversation in an unexpected direction, one that touches on her deceased father.

He asks her:

"Did you love your father?"

Ellie doesn't hesitate.

"Yes."

Palmer pauses. Then he asks a single, disarming question:

"Prove it."

The room goes quiet.

Ellie is caught off guard–not because the question is hostile, but because it exposes a limit she cannot escape. She loved her father deeply. That love shaped her life–her curiosity, her resilience, her sense of direction. It is one of the most real forces she has ever known.

And yet–

There is no instrument to measure it.
No formula to quantify it.
No experiment to reproduce it on demand.

She cannot prove it.

In that moment, the movie makes its point with remarkable clarity: some of the most powerful truths in human life are not accessible through hard, objective measurement–yet they guide our choices, anchor our values, and explain our behavior far more than any data ever could.

Love.
Meaning.
Soul.

They are not irrational.
They are not imaginary.
They are simply non-measurable.

And still, they are real.

I know that Soul-Based Regenerative Leadership will resonate deeply with some–and be dismissed by others as "touchy-feely." That is fine. I do not expect it to be embraced by everyone. I have seen it work under real business pressure. And in a world conditioned to demand proof and strict evidence before adopting any idea, skepticism is understandable.

The truth is that we now live in a world that moves too fast–and changes too deeply–for proof to always come first. By the time something is fully measured, validated, and institutionalized, it is often already too late.

So instead of asking, *"Is this bulletproof?"*
I suggest we ask a different question:

Is this true enough to bet on?
This is where the conversation about leadership in the AI era truly begins–not in certainty, but in a different state of possibility.

The Organizational Crossing

The shift into the AI era will not occur in a controlled environment.

Organizations will not pause operations, redesign themselves, and relaunch with a new model. They will have to cross while still delivering results, meeting commitments, and sustaining daily work.

This is what makes the transition uniquely difficult.

The old performance logic will continue to operate even as it loses effectiveness. Targets will still be set. Plans will still be executed. Expectations will still rise. Yet the mechanisms that once made performance predictable will no longer produce the same outcomes.

Organizations will find themselves operating between two worlds.

In one world, the inertia of performance is extracted through familiar levers: pressure, optimization, and control. In the other, a new performance logic depends increasingly on human capacity, alignment, and meaning – resources that cannot be commanded into existence.

Both worlds will coexist for a time.

Leaders will feel the tension. Effort will increase while results lag. This is not failure. It is a signal that the underlying logic of performance is changing.

The crossing requires organizations to recognize that intelligence is no longer the scarce resource it once was. As AI scales knowledge and execution, the limiting factor shifts from capability to human capacity – from what the organization knows to what its people are able to sustain, interpret, and bring forward over time.

Value creation will therefore depend less on extracting more from people and more on enabling them to operate from their best selves.

There is, however, an encouraging aspect to this shift.

When people begin to experience that their contribution truly matters – that their judgment, perspective, and effort make a visible difference – energy returns faster than expected. Engagement stops being something that must be managed and becomes something that emerges on its own.

The heavier burden, therefore falls on leaders

They must initiate change while still operating within systems designed for a different era. They are asked to redesign the logic of performance while the old one still lingers.

This is the true weight of the organizational crossing.

From Knowledge Workers to Sentient Workers

For decades, we have described decision-makers as knowledge workers. The term was coined by Peter Drucker in his 1959 book *Landmarks of Tomorrow* to distinguish workers who create value primarily through thinking and decision-making rather than through manual labor.

Knowledge workers relied on information, judgment, and expertise–acquired through education and experience–as their primary means of contribution. Productivity was no longer driven by physical effort, but by the quality of thinking. Knowledge, not labor, became the dominant source of economic value.

Drucker was right.

For decades, knowledge workers powered growth, innovation, and competitive advantage. Organizations invested heavily in skills, expertise, and intellectual capital as the engines of performance.

But the conditions that made knowledge work distinctive are now changing.

In the AI era, knowledge itself is no longer scarce. Intelligence can be accessed, augmented, and scaled at unprecedented speed. What once differentiated the best thinkers is increasingly available to anyone with the right tools.

As knowledge becomes abundant, a new question emerges:

What remains uniquely human?

The answer is not more information, faster analysis, or superior algorithms. It is the capacity to experience meaning, exercise judgment in uncertainty, embody values, and act from inner coherence.

This marks a shift from knowledge workers to what I call **sentient workers**–individuals whose value comes not only from what they know, but from how they sense, interpret, and respond to complexity.

Sentient workers bring not just skills, but awareness.
Not just intelligence, but essence.
They bring their soul to their work.

And in the AI era, that difference changes everything.

The Act of Centering

Leading beyond intelligence does not mean choosing between technology and humanity. It means becoming fluent in moving along a continuum that organizations now inhabit every day:

Non-Human-Centered Work ⟵⟶ Human-Centered Work
Automation, standardization, and optimization sit on one end.
Judgment, meaning, coherence, and soul sit on the other.

Leadership in the AI era does not live at a fixed point on this continuum.
It requires discernment in motion.

When leaders face complex or high-stakes situations, their first task is not to decide what to do. It is to discern **where to stand.**

Some aspects of a problem can–and should–be addressed through non-human-centered approaches. Automation can remove friction. Standardization can increase reliability. Optimization can improve execution.

But other aspects demand a deliberate shift into the human-centered domain–where meaning must be interpreted, values weighed, trade-offs held, and judgment exercised in uncertainty.

The leadership failure of our time is not the overuse of technology.
It is failing to recognize when execution alone no longer works.

As execution-driven activities become increasingly automated and standardized, the human-centered stance does not disappear. It becomes more decisive.

Effective leadership today is the ability to center deliberately–to move along this continuum in real time and amplify the human precisely where execution alone no longer works.

This is not a technical skill.
It is a leadership stance.
And it is the stance that defines leadership beyond intelligence.

Reader Takeaways

1. **AI amplifies intelligence, but it does not define leadership.**
   As intelligence becomes abundant and commoditized, leadership differentiation shifts from what we know and execute to how we provide judgment, direction, and meaning.

2. **The true human frontier is not technological–it is the soul.**
   AI can optimize decisions, but it cannot originate meaning, values, or moral accountability. These remain uniquely human and increasingly decisive.
3. **Human value is shifting from knowing to sensing.**
   The AI era marks a transition from knowledge workers to sentient workers–those whose contribution comes from awareness, judgment in uncertainty, and inner coherence, not expertise alone.
4. **Performance must move from extraction to regeneration.**
   Control, pressure, and intelligence no longer sustain results. Regenerative performance depends on human capacity, alignment, and meaning.
5. **Leadership beyond intelligence is a stance, not a skill.**
   Effective leaders learn to move deliberately between automation and human judgment–centering where execution alone no longer works. This is the essence of Soul-Based Regenerative Leadership and the Soul Edge of the AI era.

CHAPTER 13

The Systemic Crossing: The New Performance Logic

"You shouldn't have said that. It sends the wrong message."

That was the feedback I received from several members of the Executive Committee I was part of after a global Town Hall.

Nearly 5,000 employees had joined the session. We presented the company's results, reviewed progress on strategic initiatives, and opened the floor for questions. As often happens in large organizations, many of the questions gravitated toward HR topics–promotions, development opportunities, inclusion, and performance feedback.

One question was directed to me in my role as CHRO. An employee described a recent feedback conversation with their manager:

"He was clear in his message, but I didn't agree with his assessment. I didn't want to come across as defensive, so I listened carefully. To show that I understood, I even repeated his words back to him. But I can't fix something I don't agree with. What should I do?"

I paused and responded with a personal story.

"There was a time when my own boss told me that I lacked certain skills–skills I knew I had. I left that session frustrated, convinced he was wrong. Later, after reflecting, I considered another possibility: perhaps I had not demonstrated those skills clearly enough in our interactions. So I made a deliberate effort to bring them forward more visibly."

Then I added:

"You may know you have something. But it also has to be seen."

And I continued:

"And if, after making it visible, it still doesn't come through, perhaps you and your boss simply hold different standards. It's possible to disagree, as long as you are open about it."

The Town Hall concluded, and the Executive Committee gathered for a debrief to align on next steps. After covering the remaining topics, a few colleagues returned to my response.

"You shouldn't have said that."

"You can't suggest that a manager's assessment might be wrong. How is a leader supposed to deliver results if subordinates question their judgment?"

I tried to explain my position, but it soon became clear that we were not debating feedback techniques.

We were defending a system.

The organization's performance model was anchored in control and the steady application of pressure. Questioning a manager's judgment–even respectfully–was perceived as introducing instability. The concern was not about communication. It was structural.

The system was built around a particular performance logic, and that logic had delivered results for many years. From that vantage point, their concern was understandable. I did not agree, but I could see their reasoning. The Town Hall exchange did not create the tension; it exposed it.

You have probably experienced something similar–receiving feedback you did not fully agree with, or nodding along with a decision you were not entirely on board with, simply to move forward.

These moments rarely make headlines. Yet they quietly reveal the logic that governs how performance operates inside the system.

This episode happened a few years ago. At the time, the system still held. Today, however, the context calls for a different way of thinking about performance.

We are living through a systemic crossing–not merely a shift in leadership style or communication practices, but a deeper transition in how performance itself is generated in the age of AI.

What happens when performance depends on expanding human capacity rather than reinforcing control or pressure?

This is not an abstract question. It shapes how roles are designed, how decisions are made, and how value is created.

That is the systemic crossing before us.

The Invisible Equation

In his well-known commencement address *This Is Water*, David Foster Wallace opens with a simple story. Two young fish are swimming along when they pass an older fish, who nods at them and says, "Morning, boys. How's the water?" The two younger fish continue swimming for a while, until one looks at the other and asks, "What the hell is water?"

We are often like those fish.

We live inside systems so deeply embedded in our daily routines that we rarely stop to examine them. They feel natural, inevitable–just the way things are.

Think about how you operate within your organization. You commit to results. You follow established processes and procedures. You work within a structure that directs your actions and defines your responsibilities. If you execute well–if you deliver with precision and operational excellence–you are rewarded. The formula repeats. Over time, it becomes self-evident: perform, deliver, repeat.

When this method works, organizations achieve what many call their "business edge"–the distinctive capability that allows them to differentiate in the market, create value, and grow. The organizational engine is built for efficiency. Its visible components are familiar: purchasing, production, logistics, distribution, sales, marketing, administration, technology. Each part has a defined function, and each contributes to the whole.

But beneath these visible components lies a quieter architecture–an implicit equation that governs how performance is understood and produced.

It typically unfolds in a familiar sequence:

- Define the job.
- Identify the required skills.
- Place the person into the role.
- Optimize performance.

Skills become the entry point. Roles become containers. The system coordinates the parts.

If you have participated in a performance review or been positioned within a nine-box grid, you know this logic firsthand. Skill-based talent planning becomes the platform for development, promotion, succession–and sometimes exit. The system is not accidental. It is designed to ensure that performance is achieved with consistency and predictability.

For many years, this logic worked remarkably well.

It was fueled by three levers: control, codified intelligence, and pressure. Clear authority ensured alignment. Proprietary knowledge created advantage. Sustained pressure maintained momentum. Together, they formed the kernel of modern performance.

In recent years, however, subtle cracks have begun to appear. Burnout has become commonplace. Disengagement is no longer an exception. Conversations about work-life balance signal a deeper fatigue within the system.

In response, organizations have attempted to humanize performance. Engagement programs were introduced. Purpose statements were articulated. Value frameworks were refined and communicated. These efforts were well-intentioned–and in many cases necessary. Yet they often functioned more like patches than structural solutions.

The underlying architecture remained largely unchanged.

We attempted to humanize performance–but we did not redesign it.

And as long as the environment remained relatively stable, the system held. After all, if it wasn't broken, why fix it?

But AI Changes Everything

Now the context is fundamentally different.
We saw it coming – but assumed we still had time.

When artificial intelligence first entered the conversation, it felt distant–an emerging technology that would take years, perhaps decades, to meaningfully reshape how organizations operate. That assumption proved wrong.

AI has moved through business quickly. Companies across industries are investing heavily, experimenting rapidly, and rethinking how work will be done. What once felt theoretical is now operational.

We may not yet know exactly where this transformation will lead, but we can already see its direction.

Repetitive tasks are being automated. Analytical work is being accelerated. Agentic systems are beginning to handle functional tasks that, until recently, required human expertise. What was once scarce–codified knowledge–is becoming widely available and accessible.

Three shifts are becoming clear:

- Codified intelligence is no longer a reliable differentiator.
- Execution can be scaled faster than ever.
- Skills, once developed internally or sourced externally, can increasingly be accessed–or "borrowed"–through technology.

This directly affects the core of the old performance logic.

If performance was fueled by control, codified intelligence, and pressure, one of those pillars is now fundamentally altered.

When intelligence becomes widely available, it no longer guarantees advantage. The equation begins to lose its balance.

The other two levers–control and pressure–had already shown signs of strain through burnout and disengagement. But AI challenges something even deeper: the assumption that proprietary knowledge is the main driver of value creation.

When that assumption weakens, the logic beneath the system must be reconsidered.

If codified intelligence becomes abundant, what remains scarce?

Human capacity.

Not just what people know–but how they think, adapt, collaborate, and respond to change. How we cultivate, align, and expand human capacity becomes central to performance itself.

This shift does not affect organizations alone. An entire ecosystem emerged to support the old performance equation. Business schools, consulting firms, executive education programs, leadership institutions, executive search firms, assessment tools, workforce planning models, and performance management systems were all designed around a common assumption: that results are generated through control, codified intelligence, and optimization.

As organizations move toward regenerative performance, this ecosystem must evolve as well. Those that remain focused primarily on optimizing the old equation risk becoming less relevant. Those that help organizations expand human capacity will become increasingly valuable.

Skills Are Necessary, but Not Sufficient

In the old performance equation, we tended to see people primarily through the lens of skills. If you brought the required expertise to the table, you could produce results. Skills defined contribution. Development meant acquiring more of them. Performance meant applying them effectively.

The guiding logic for organizations was simple: secure the right person (based on skills), place them in the right role (where those skills were needed), and execute with precision. That approach made sense in a world where knowledge was scarce and roles were relatively stable.

But in the new performance equation, skills are no longer the whole story.

They remain necessary. Without competence, there is no performance. But skills alone do not determine results in a context that is constantly shifting–especially when many skills can now be automated or augmented by technology.

Today, a person is more than a set of functional capabilities. Each individual brings judgment, adaptability, resilience, creativity, relational intelligence, and a unique way of interpreting situations. These human qualities shape how skills are applied–and how they evolve as circumstances change.

Skills become an entry point, not the endpoint. They must remain fluid, continuously updated as technology advances and roles transform. The shift is subtle but decisive.

Instead of asking,
"What skills does this role require?"

We begin to ask,
"What human qualities can this person bring to the role?"

The focus moves from fitting people into predefined slots to designing roles where human capacity can grow. This is the Soul-Based approach.

Skills are necessary–but they are no longer sufficient to define performance.

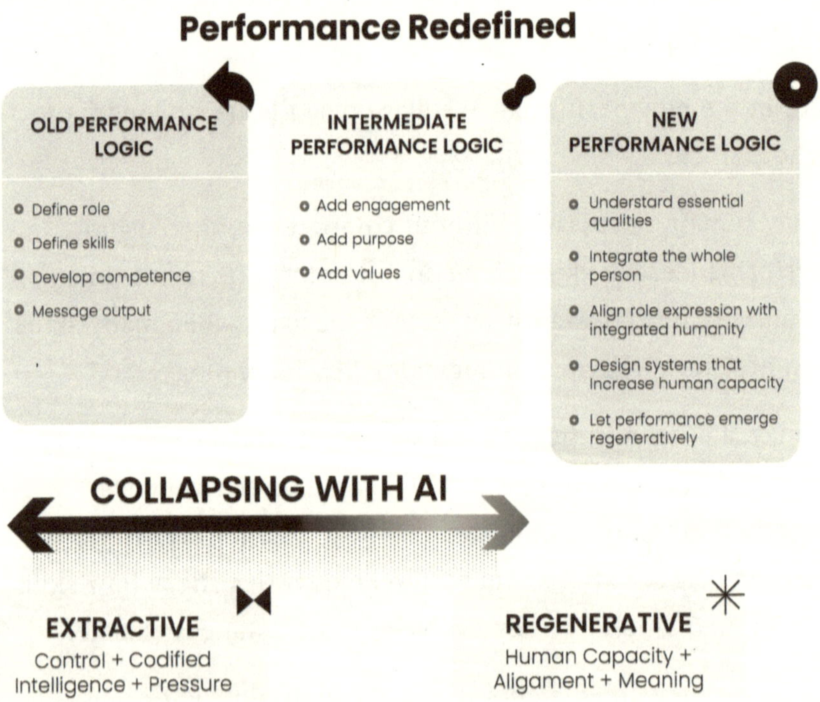

The Systemic Crossing

Regenerative performance should not be mistaken for a softer or less rigorous approach. It introduces a different kind of rigor.

Operational excellence still matters.
Efficiency still matters.
Measurement still matters.

What changes are the levers that drive performance.

Human capacity expands output.
Alignment reduces friction.
Meaning sustains energy.
Integrated humanity becomes the multiplier.

If regenerative performance becomes the outcome, the system reorganizes around it. What organizations look for in people, how they develop them, how they evaluate contribution, and how they shape culture all begin to shift. Often without explicitly announcing it, the organization moves from optimizing for extraction to optimizing for regeneration.

The individual is no longer seen as a collection of skills, but as a multifaceted person bringing a unique combination of qualities. Skills, mindset, soul signature, and the capacity to draw from a deeper source of insight together form the foundation of human capacity, alignment, and meaning.

Regenerative leadership is not about making performance more humane.

It is about redefining performance so that integrated humanity becomes its anchor.

In the old model, organizations pursued their "business edge" – the distinctive capability that allowed them to differentiate, compete, and create value.
In the new model, the search goes deeper.

The system itself becomes the capability that creates advantage.
As a leader, you will confront moments when the path forward is unclear and the answers do not come. This is no longer the exception – it is the new leadership environment.

You are not expected to have all the answers.
You are expected to create the conditions in which they can emerge – from the collective contribution, essential qualities, and human capacity of your people.

How you shape that space will determine whether the system depletes itself through pressure and control, or regenerates from within.

Organizations must therefore discover their Soul Edge – the point at which human capacity, alignment, and meaning converge to generate sustainable growth and enduring value.

The new business edge is the Soul Edge.
That is the systemic crossing.

Reader Takeaways

1. **Every organization operates from an implicit equation of performance.**
   Whether recognized or not, there is a theory about how results are produced that shapes roles, decisions, incentives, and behavior across the system.
2. **The traditional performance model is reaching its limits.**
   Control, proprietary knowledge, and pressure once drove results, but they now generate burnout, disengagement, and diminishing returns in a rapidly changing environment.
3. **AI is shifting the source of competitive advantage.**
   As intelligence becomes abundant and execution scalable, the decisive factor in performance shifts from knowledge to human capacity.
4. **Human capacity is driven by new performance levers.**
   Capacity expands output, alignment reduces friction, meaning sustains energy, and integrated humanity becomes the multiplier that enables sustainable results.
5. **The systemic crossing reshapes the architecture of performance.**
   As organizations move from extractive to regenerative performance, the systems that support talent development,

evaluation, and leadership must evolve as well – inside organizations and across the broader management ecosystem.

6. **The new business edge is the Soul Edge.**
   In the emerging landscape, sustainable value comes from expanding human capacity and reaching the human frontier.

AUTHOR'S NOTE

Leadership for Mortals – A Human Way to Lead in the Age of Intelligence

Much of today's leadership content leaves me with a knot in my stomach.

It always begins the same way – a tightening right below the rib cage.
It shows up whenever I hear yet another leadership sound bite: an overused slogan disguised as wisdom, or a hyper-complex framework delivered with academic flair and little regard for real life.

It's a physical reaction. I can't help it.

Why does leadership have to be like this?

Why does it keep getting pulled toward two equally frustrating extremes?

After more than three decades inside small and large, global, complex organizations – leading businesses, cultures, and talent systems – I have never met a human being who actually leads from either of those poles. And yet, that is what dominates the leadership "menu" today.

On one side sits the fast-food version of leadership: shallow, formulaic, superficially satisfying. Motivational slogans. Copy-paste models. Bite-sized tips engineered for clicks, keynotes, and workshops. They taste good for a moment but leave you undernourished and empty.

On the other extreme is pretentious gourmet leadership: dense, conceptual, high-theory frameworks served by elite consultants and academics. Elegant plates. Beautiful diagrams. Nearly impossible to apply in real life. Leadership so abstract it only works on whiteboards – the kind of restaurant where you don't recognize the silverware, the portions are microscopic, and you can't pronounce the name of the dish.

Between empty carbs and haute-cuisine ego trips, real leaders are left hungry.

And hungry leaders don't grow.
Hungry organizations don't regenerate.
Hungry people don't rise to their best selves.

Maybe you've felt this frustration too.

Maybe you've tried to nourish your leadership and kept finding options that were either too shallow to matter or too pretentious to use.

The AI era is now forcing us to confront this gap directly. The way we have driven performance for decades is becoming a constraint, not a solution. Human capacity is being eroded faster than it is renewed, and leaders need something accessible – yet deep enough – to meet what this moment demands.

That is why I wrote *The Soul Edge.*

Soul-Based Regenerative Leadership is meant to be that place.

It begins with awareness, builds through skills and mindset, deepens into the discovery of one's soul signature, and ultimately opens into a higher field of insight.

Not fast food.
Not fine dining reserved for special occasions.
But leadership you can return to – every day.
Nourishing enough for longevity.
Simple enough for real life.
Deep enough to transform you and others.

This book was written for leaders who are accountable for the performance of their teams, who feel stuck in systems delivering diminishing returns, and who sense that AI will inevitably push them to explore a different way to lead.

If you find yourself seeking:

- a way to close the performance gap without further exhausting people
- a leadership approach that regenerates human capacity
- language and structure to redesign how work and leadership actually operate
- a coherent model that aligns performance, people, and the ability to adapt as conditions change

then this journey was meant for you.

But if you step back now, at the end of the journey, something else becomes visible.

What we have explored is not merely a leadership approach.

It is part of a broader shift in how organizations will create value in the age of intelligence.

The New Performance Logic

Regenerative Performance is the next evolution of how organizations produce results when intelligence is no longer scarce and pressure no longer works.

Extractive performance models – built on control, pressure, and ever-increasing demands on people – are reaching their human and systemic limits. They can still deliver results for a time, but at the cost of depletion: energy, creativity, trust, and ultimately long-term performance.

Regenerative performance operates differently.

Human capacity expands output.
Alignment reduces friction.
Meaning sustains energy.
Integrated humanity becomes the multiplier.

Soul-Based Regenerative Leadership is the leadership system that makes this possible.

It is not merely a mindset or a philosophy, but an operating logic that can be applied at the level of individuals, teams, and organizations.

Performance regeneration restores the human capacity on which durable performance depends.

The task ahead is therefore twofold: to renew the organizational systems through which work gets done and to renew the human beings who lead and sustain them.

This requires a new performance architecture – one that treats regeneration not as a soft alternative, but as a strategic necessity.

The real question before us is simple:

What does performance truly mean in the age of intelligence, and how should it be achieved?

Where This Leaves You

As much as any author hopes to contribute to significant change – and that is certainly my case – I know I can only go so far. The real work belongs to the leaders who recognize the need and choose to act on it.

If that is you, then consider this an invitation to the crossing.

Leadership is for mortals. That means there is no perfect formula, no bulletproof method, and no guarantee of success. But it also means leadership is accessible to those willing to grow, reflect, and lead with humanity under pressure.

Performance regeneration is the overarching goal, and no artificial-intelligence shortcut will produce it. It is distinctly human work – done by humans, for humans.

Leadership for mortals – imperfect, accountable, learning in real time – may turn out to be exactly what this era requires.

Leadership capable of regenerating rather than depleting.

Leadership that remembers performance is not only about what organizations extract from people, but about what people become in the process.

The real work begins now.

In every decision you make, choose regeneration over extraction.

I wish you the very best on your journey to lead as a human being first – committed to leaving people and systems closer to their human frontier than you found them.

APPENDICES

The Regenerative System

The following appendices bring together the core elements of the shift explored in this book.

They are not designed as supplementary material, but as practical lenses to understand, apply, and communicate the transition toward regenerative performance.

Each appendix offers a different entry point into the system–from navigation, to language, to visual models–so that the ideas in this book can be translated into action.

APPENDIX I

Navigating the Shift

A practical guide to transitioning toward regenerative performance

As I was writing this book, one thing struck me as odd–though perhaps it is normal whenever a paradigm begins to shift. The old paradigm often appears alive and well, even when its limitations and shortcomings are increasingly visible. It still has committed followers, established practices, and a long record of success. Yet the reasons for change remain elusive until the system itself reaches a tipping point.

In podcasts, business media, and leadership forums, I still hear stories of CEOs who engineer dramatic turnarounds by doubling down on extractive performance. They are celebrated, almost mythologized. Their stories are showcased as models of decisive leadership, and their methods are widely recommended.

Yet when you speak with people inside these organizations–especially those working in the middle of the system–you often hear a very different story. Beneath the headlines of success lies a growing sense of human depletion. Long before the disruptive force of artificial intelligence entered the scene, many organizations were already operating close to exhaustion. In that sense, the system has been a ticking bomb.

Artificial intelligence is not the origin of this tension–but it may well be the accelerant that exposes it.

As intelligence becomes widely accessible and increasingly embedded in business processes, the traditional sources of advantage begin to erode. What once differentiated organizations gradually becomes table stakes. The performance model that has dominated for decades does not disappear overnight, but its limitations are becoming increasingly difficult to ignore.

This appendix offers a way to understand what may come next. It brings together the ideas explored throughout the book into a conceptual map of the transition from extractive performance to regenerative performance.

Here, I summarize the stages organizations are likely to navigate as this transition unfolds. The shift is already underway, and it may not take long before the value-creation system we have relied on for decades begins to reveal deeper dysfunctions.

The important question, therefore, is not whether the transition will occur, but how leaders can begin preparing for it.

To clarify this challenge, the appendix presents two complementary perspectives.

The first is a **descriptive model** that outlines how the shift from extractive performance to regenerative performance may unfold across industries and organizations.

The second is a **developmental path** that describes how leaders can begin preparing for this transition today.

These two perspectives operate simultaneously. While the external system evolves over time, leaders must begin developing the capabilities required for the next era before the shift becomes fully visible.

The system may change gradually, but preparation cannot wait. Leaders must continue operating within existing structures even as they begin cultivating the capacities required for what comes next.

Descriptive Model – The Performance Transition

A descriptive view of how the system evolves

Stage 1 – Extraction

The dominant performance logic
For most of the modern business era, organizations have operated under a performance equation that relies on scaling intelligence and execution.

Performance is generated through:

Control + Codified Intelligence + Pressure

Leaders design systems that translate knowledge into processes, playbooks, and management structures. Control mechanisms ensure alignment, while performance pressure drives output.

Within this system, leadership advantage is associated with the ability to generate superior analysis, formulate strategy, and mobilize execution.

The prevailing leadership archetype is the **Heroic Cognitive Leader**–a leader whose authority comes from insight, expertise, and the capacity to drive results.

This model has delivered remarkable growth and efficiency for decades. Even when signs of strain appear–burnout, fragmentation, and diminishing returns–they are often interpreted as the unavoidable cost of high performance rather than a limitation of the system itself.

Stage 2 - AI Amplification

The illusion of improvement

As artificial intelligence becomes increasingly integrated into business operations, many organizations initially experience noticeable improvements in speed and efficiency.

AI enhances the existing system by accelerating tasks such as:

- analysis and forecasting
- marketing and pricing optimization
- operational planning
- customer service and product development

These gains often reinforce the belief that the current performance model simply needs better tools.

In this phase, organizations encounter what can be called the **Rational Dilemma**: the improvements created by AI make it appear that the existing performance model is becoming stronger, precisely at the moment when its foundations are beginning to erode.

Because results initially improve, leaders conclude that the system is working exactly as intended. The rational response, therefore, is to double down on the same logic that has driven performance for decades: increase efficiency, tighten control, and push productivity further.

Artificial intelligence seems to validate the extractive equation rather than challenge it.

Yet beneath these improvements, a deeper structural shift begins to take place.

Artificial intelligence dramatically expands access to codified intelligence. Capabilities that once differentiated organizations–advanced analytics, optimization models, and forecasting techniques–become increasingly accessible across industries.

As these capabilities spread, the sources of competitive advantage that once distinguished organizations begin to erode. Intelligence, once scarce and valuable, becomes progressively abundant.

In this stage, the extractive system is not replaced. Instead, it is temporarily amplified, even as the foundations of its advantage begin to weaken.

The system becomes faster, more optimized, and more intelligent.

But the logic that once generated differentiation is quietly dissolving.

Stage 3 - System Fracture

When intelligence becomes table stakes
As artificial intelligence spreads across industries, algorithms increasingly influence–or directly automate–many of the decisions that once depended on human expertise.

These include:

- pricing strategies
- marketing optimization
- supply chain planning
- financial forecasting and risk modeling

Because many organizations adopt similar technologies and datasets, these decisions become progressively optimized across the market.

What once produced competitive advantage gradually becomes table stakes.

Organizations begin making similar strategic and operational moves. Pricing strategies converge. Marketing optimization yields comparable results. Operational efficiencies level out across competitors.

Companies become faster, more optimized, and more intelligent–yet differentiation becomes harder to sustain.

At this stage, the system begins to encounter a deeper structural tension.

The extractive performance model continues to push for greater efficiency, tighter control, and stronger execution. But as optimization spreads across industries, these improvements no longer generate the same strategic advantage.

Instead, organizations begin experiencing what might be called **Systemic Suffocation.**

The system becomes so optimized around control, codified intelligence, and performance pressure that it gradually loses its capacity to renew itself.

Innovation slows because experimentation appears inefficient.

People experience increasing pressure and fatigue.

Organizations struggle to rethink their assumptions because the system rewards execution rather than exploration.

The organization becomes better at doing what it already knows how to do.

But it becomes less capable of discovering what it does not yet know.

At this point, a deeper realization begins to emerge.

When intelligence becomes widely available, advantage must come from elsewhere.

The frontier of differentiation shifts toward distinctly human capabilities, including:

- judgment under uncertainty
- creativity and imagination
- the integration of diverse perspectives
- ethical discernment
- the ability to generate trust, meaning, and coherence

The challenge is no longer producing intelligence.

It is activating the human qualities that intelligence alone cannot replicate.

Stage 4 – Regenerative Shift

The emergence of a new performance logic

As the limitations of extractive performance become increasingly visible, some organizations begin to recognize that the challenge is no longer simply improving the system.

The performance equation itself must evolve.

For decades, performance was generated primarily through the scaling of intelligence and execution–through control, codified intelligence, and pressure. But when intelligence becomes widely accessible and optimization spreads across industries, these mechanisms alone are no longer sufficient to sustain advantage.

At this stage, a new leadership logic begins to emerge.

Instead of relying primarily on extracting more output from systems and people, organizations begin redesigning their operating model around a different equation:

Regenerative Performance = Human Capacity + Alignment + Meaning

In this new logic:

- **Human capacity** refers to the activation of essential human qualities–such as clarity, curiosity, communication, gratitude, service, creativity, empathy, wisdom, courage, and harmony.
- **Alignment** ensures internal coherence, allowing these essential qualities to reinforce one another and amplify what is genuinely human across the organization.
- **Meaning** generates the transformational energy that moves the organization forward as people increasingly become what they are meant to be.

Within this emerging paradigm, leadership turns toward what can be called the **Human Frontier**.

As intelligence becomes widely accessible, the frontier of performance moves inward, toward the cultivation of the human qualities that enable organizations to learn, adapt, and renew themselves.

This shift also transforms the dominant leadership archetype.

Instead of the **Heroic Cognitive Leader**, whose authority derives primarily from expertise and analytical insight, organizations increasingly require leaders capable of cultivating coherence, awakening these essential human qualities, and guiding their organizations through continuous renewal.

This emerging archetype is the **Soul-Based Regenerative Leader**.

Organizations that successfully navigate this transition avoid the false trade-off between efficiency and human capacity. They recognize that efficiency is no longer a distinctive advantage and instead cultivate the human qualities that regenerate performance.

Developmental Path – Leadership Preparation

A developmental view of how leaders prepare for the transition

While the system transition described above unfolds unevenly across industries, leaders do not have the luxury of waiting for the shift to become obvious.

Preparation must begin earlier.

The path described throughout this book unfolds through three interconnected movements.

Movement 1 – Clearing the Field

(Part I of the book)

The shift toward regenerative performance begins with a deceptively simple act: **making space**.

Over time, leadership has accumulated layers of frameworks, models, and prescriptions that once promised clarity but now produce noise. Instead of coherence, leaders experience overload and fragmentation.

The first step is therefore **distillation**.

Leaders must simplify their practice by releasing what no longer serves:

- excess frameworks and management tools
- unnecessary complexity in decision processes
- the cultural imperative of doing more with less

This shift is not destructive. It is clarifying.

By focusing on fewer actions that truly matter, leaders create the space required for regenerative leadership to emerge.

Movement 2 – Cultivating Human Capacity

(Part II of the book)

With the field cleared, leaders begin developing the capabilities that will define advantage in the AI era.

The **Soul-Based Regenerative Leadership framework (SBRL)** offers a regenerative path that unfolds through four movements:

1. **Mastering the Tools (Skills)** – acting with clarity and competence
2. **Expanding the Horizon (Mindset)** – thinking differently and embracing learning
3. **Discovering the Essence (Soul Signature)** – anchoring leadership in essential qualities
4. **Opening to the Higher Field (Receiving)** – accessing creativity, insight, and inspiration

These movements reconnect the full arc of leadership:
doing, thinking, being, and transcending.

At the center of this process lies a foundational capability: **awareness**, the red thread that shapes the quality of every decision and interaction.

Movement 3 - Preparing for the Crossings

(Part III of the book)
As leaders deepen their capacity, they begin preparing for the transitions required in the AI era.

These transitions unfold through three crossings.

The leader's crossing – from hero to steward

Leaders release an identity built on certainty, answers, and control, and embrace a role centered on cultivating awareness, connection, and collective intelligence.

The organizational crossing – leading beyond intelligence

Organizations move from relying primarily on knowledge workers to cultivating sentient workers–individuals whose contribution comes not only from what they know, but from how they sense, interpret, and respond to complexity.

As intelligence becomes abundant, performance shifts from extraction through pressure and control to regeneration through human capacity, alignment, and meaning.

The systemic crossing – the emergence of a new performance logic

Organizations move beyond a system designed to match skills to roles and optimize execution. As intelligence becomes abundant, skills remain necessary but no longer sufficient.

Instead of asking "*What skills does this role require?*" leaders begin asking "*What human qualities can this person bring to the role?*"–shifting performance toward understanding the range of human qualities individuals bring and assessing the integrated spectrum through which those qualities express themselves.

Bringing the Two Models Together

The system transition and the leadership path unfold simultaneously.

Organizations will continue operating within extractive systems even as those systems begin to lose their effectiveness. Leaders, therefore, face a genuine dilemma.

They cannot stop the plane to redesign it. They must fly it while fixing it.

Waiting for certainty before acting may feel prudent, but in a world moving faster than certainty can arrive, leadership requires a different stance.

The essential question becomes:

Is this true enough to bet on?
Those who begin preparing early–by simplifying leadership, cultivating human capacity, and preparing for the crossings ahead–will be better positioned to navigate the shift toward regenerative performance.

A Final Reflection - The Limits of the Extractive System

For decades, leaders have been right to trust the extractive performance model. It created extraordinary value and helped build many of the organizations that shape our world today. In many ways, the system worked exactly as intended.

This also means that the leaders who promoted it were not acting out of disregard for people. The erosion of human vitality that we now observe was rarely the product of deliberate design. It was the consequence of a system optimized for efficiency, scale, and output.

Those who championed this model believed that its benefits outweighed its costs. Even when the human toll became visible, it was often interpreted as the unavoidable price of progress–the "casualties of war" in the pursuit of growth and competitiveness.

Seen from that perspective, the extractive system has done exactly what it was designed to do.

But leaders must realize that every system eventually encounters the limits of its own logic.

As artificial intelligence accelerates the availability of intelligence itself, the foundations of the extractive equation begin to shift. What once created advantage becomes increasingly accessible, and the sources of differentiation begin to move elsewhere.

This is why regeneration is not merely a leadership preference–it is becoming a structural necessity.

The old system cannot simply be patched. It fulfilled its role, but the conditions that sustained it are changing. A new perspective on performance is emerging–one that places **regeneration at the center of value creation.**

In that sense, the transition described in this book is not a rejection of the past, but the natural evolution of a system that has reached the limits of what it can produce.

The next chapter of leadership will unfold at a different frontier.
The Human Frontier.

And the competitive advantage will no longer come from codified intelligence.
It will come from a different edge.

The Soul Edge.

APPENDIX II

The Language of Regenerative Leadership

Key concepts to think, speak, and lead differently

The concepts below form the core vocabulary of regenerative leadership and the performance logic introduced in this book. Together, they provide a language for understanding the transition from extractive to regenerative performance in the age of artificial intelligence.

I. The New Leadership Frame

1. Soul Edge

The **Soul Edge** is the distinctive leadership advantage that emerges when individuals and organizations consciously cultivate their essential qualities and inner coherence. Unlike skills, knowledge, or technology, it cannot be copied, automated, or outsourced. It represents the defining human differentiator in the AI era.

2. Soul-Based Regenerative Leadership (SBRL)

A leadership approach that places human vitality, meaning, and inner coherence at the center of sustainable performance. SBRL moves beyond control and optimization toward renewal and long-term value creation by leading from the inside out. It unfolds through four developmental movements–**Skills, Mindset, Essence, and Higher Field**–anchored in awareness.

3. Human Frontier

The emerging domain of value creation where uniquely human capacities–awareness, judgment, ethics, creativity, and meaning–become decisive. As technology advances and intelligence becomes abundant, the human frontier expands rather than contracts. It is the inner territory where sustainable advantage increasingly resides.

4. Extractive Performance

A performance model rooted in control, codified intelligence, and sustained pressure. It generates results by maximizing output from skills and systems, often at the expense of human energy, coherence, and long-term vitality. While it can produce short-term gains, it ultimately leads to burnout, disengagement, and diminishing returns.

5. Regenerative Performance

A performance model centered on expanding human capacity rather than consuming it. It produces sustainable results through the alignment of human capacity, shared meaning, and inner coherence. In regenerative performance, output grows because people grow; energy is renewed rather than depleted.

II. Making Space

6. Push vs Pull

Two fundamental leadership energies.
Push mobilizes action through authority, urgency, and control.
Pull mobilizes energy through trust, invitation, and connection.
While push may be necessary in moments of urgency, pull becomes the regenerative baseline in the AI era–unlocking creativity, initiative, and sustained engagement.

7. Trait vs Counter

A leadership polarity within the **Leadership Diamond.**
A **Trait** is a strength a leader naturally relies on (e.g., decisiveness, confidence, drive).
The **Counter** is its balancing complement (e.g., inquiry, humility, patience).
Mastery lies not in maximizing a trait but in moving fluidly between trait and counter according to what the moment requires.

8. The Leadership Landscape Map

A framework that organizes leadership activity into four quadrants based on focus (**inner vs. outer**) and intent (**prescriptive**

vs. transformative). It clarifies where organizations invest their attention and highlights the **transformative inner quadrant** as the regenerative, human-centric domain where sustainable differentiation emerges.

9. Do an Informed Less

A deliberate reduction of activity, inputs, and initiatives–guided by discernment rather than retreat. It means choosing fewer priorities with greater clarity, resisting noise and trend-chasing, and protecting attention so that what truly matters receives depth and energy.

10. Do Less, Better

A regenerative operating principle that channels effort toward a smaller number of high-leverage activities executed with excellence and coherence. Rather than multiplying initiatives, leaders refine and deepen what creates disproportionate value.

III. The New Core

11. The "I" and the "Me"

A distinction between awareness and behavior. The **"Me"** is the personality that acts, reacts, performs, and defends. The **"I"** is the observing center of clarity and choice–the inner vantage point from which leaders notice patterns and respond intentionally rather than automatically.

12. Skills

The capabilities and competencies that enable execution and performance. Skills represent the **"doing"** dimension of leadership development. In the regenerative model, skills remain necessary but are no longer sufficient; they are strengthened and guided by mindset, essence, and alignment.

13. The Inner Reset

A conscious interruption of automatic reactions, limiting beliefs, or inherited patterns. The **Inner Reset** realigns perception, energy, and intention so leaders can respond from awareness rather than habit. It allows performance to evolve by shifting internal constraints rather than merely adjusting external behavior.

14. Essential Qualities

Innate human attributes–such as courage, clarity, compassion, integrity, creativity, and service–that express the core of who we are. They are not acquired competencies but awakened dimensions of our nature. When embodied, essential qualities generate meaning, authenticity, and the regenerative energy that sustains long-term performance.

15. Soul Signature

The unique configuration of essential qualities that forms a person's inner blueprint. A leader's soul signature shapes presence, decisions, and influence beyond role, title, or expertise. It is the distinctive pattern through which inner coherence is expressed in action–and through which enduring value is created.

16. Higher Field

A state of conscious openness beyond ego, urgency, and habitual problem-solving. When leaders loosen their cognitive grip and become receptive, insight, inspiration, and clarity emerge from inner coherence. Action then flows from awareness rather than reaction.

IV. Crossing the Human Frontier

17. Heroic Leadership

A traditional leadership model centered on certainty, control, personal expertise, and the expectation that the leader must carry the answers. While effective in stable environments, heroic leadership can limit collective intelligence and weaken long-term human capacity if not balanced by stewardship.

18. Sentient Worker

A contributor whose primary value lies not in executing tasks or processing information, but in sensing, interpreting, creating meaning, and exercising judgment. Sentient workers complement AI by bringing awareness, ethical discernment, and human insight into work.

19. Speak to the Soul, Not the Role

A leadership stance that addresses the person behind the position rather than the function they perform. It engages essential qualities, meaning, aspiration, and inner coherence–not merely tasks, titles, or performance metrics. Speaking to the soul activates energy and commitment that speaking only to the role cannot reach.

20. The New Performance Logic

The systemic shift from a performance equation based on **control, codified intelligence, and pressure** to one grounded in **human capacity, alignment, and meaning**. As AI scales intelligence and automation, competitive advantage increasingly depends on cultivating coherence, renewal, and integrated humanity within the system itself.

APPENDIX III

Visual Models of the Performance Shift

Seeing the system that drives regenerative performance

The transition from extractive to regenerative performance can be understood through three complementary lenses.

First, the shift in the performance equation itself.
Second, the structural limitations of the extractive model as it reaches its limits.
And third, the multi-layer transformation required across leadership, systems, and organizational energy.

The following visual maps bring these perspectives together.

TWO PERFORMANCE EQUATIONS

The transition from extractive performance to regenerative performance

EXTRACTIVE PERFORMANCE

Control + Codified Intelligence + Pressure

→

REGENERATIVE PERFORMANCE

Human Capacity Alignment + Meaning

EXTRACTIVE OUTPUT → **INTEGRATED HUMANITY**

EXTRACTIVE OUTPUT

- Optimization
- Efficiency
- Execution
- Performance Pressure

REGENERATIVE OUTPUT

- Adaptive capacity
- Creativity
- A living cycle of learning
- Enduring performance through renewal

The Performance Shift

SELF-INFLICTED EROSION

A self-inflicted wound: optimizing into irrelevance as advantage turns into table stakes.

SYSTEMIC ADVANTAGE

The system itself becomes the source of advantage.

The Systemic Suffocation Curve

Why the efficiency-driven model erodes

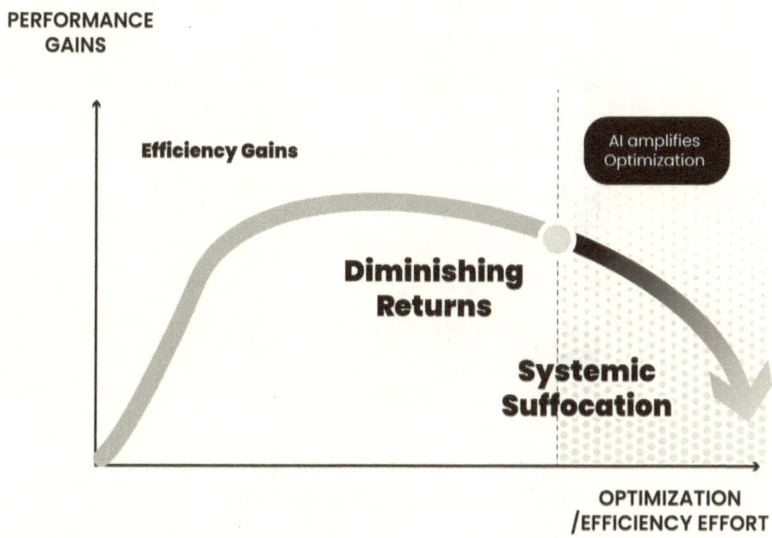

EFFICIENCY GAINS

Extractive performance initially produces strong results.

RENEWAL

Over-optimization reduces innovation, adaptability, and renewal.

THE PERFORMANCE SHIFT ARCHITECTURE

A structural view of the transition from extractive to regenerative performance across key organizational layers

	EXTRACTIVE PERFORMANCE	REGENERATIVE PERFORMANCE
Logic	Control + Codified Intelligence + Pressure	Human Capacity Alignment + Meaning
Leader stance	Heroic directive problem-solver	Steward, integrator, developer of people
Emotional energy	Fear	Care
Consciousness Mode	Reactive, automatic, ego driven	Aware, reflective, intentional
System Outcome	Efficency › Optimization › Rigidity	Adaptability › Learning › Renewal

FEAR EXTRACTS.
CARE REGENERATES.

The shift is not one model to another,
but from one energy to another.

ACKNOWLEDGEMENTS

I am convinced that we are the result of our interactions. The people who cross our path shape who we become far more than we realize at the time.

My family is the first I must thank. They have been my anchor and the reason I continue to strive, not only because I want to be there for them if they need me, but because I have always believed that deeds speak louder than words. Whatever I have built, I have built with them in mind.

Many people have placed their trust in me over the years, offering opportunities, opening doors, or simply being generous enough to have a conversation that changed the direction of my thinking. At work, with friends, and in formal education, I have been fortunate to encounter role models whose example still lives in me. I am grateful to all of them.

This book exists in its finished form because of Jane Tabachnick. She guided me through every stage of bringing a manuscript into being, from formatting to design to the many details that turn an idea into something a reader can actually hold. Her patience and her eye for what mattered shaped this book as much as any single idea inside it. I am deeply grateful.

I also want to thank Carlos Ortiz, principal at Cranea, who guided the creative process behind this book's cover and illustrations with care and a clear point of view, and Israel Ovalle, the graphic designer whose work gave form to the images that now carry so much of what this book is trying to say. The eye that appears throughout these pages began as their work before it became something readers would recognize as the heart of The Soul Edge.

There is another group that deserves acknowledgement: the authors, thinkers, and artists whose work I have read, watched, and listened to throughout my life. We live in an extraordinary moment - one in which the ideas of people who existed hundreds or even thousands of years ago are available immediately to anyone willing to seek them out. I have drawn deeply from that well in living my life and in writing this book, and I am grateful for every mind that contributed to it.

And finally, I want to thank God. I have been fortunate in ways I did not always recognize in the moment. You never know what life will bring you, and the experiences I have lived, including some I would not have chosen, have made me the person I am today. I thank God every day and trust His wisdom. I also know he has been patient with me, despite my being probably one of his more intense askers. I owe him a particular gratitude for the illumination that made this book possible, and I am genuinely excited to see where it leads. So, God, thank you from the bottom of my soul.

I have always considered myself a late bloomer. In my own mind, reaching my goals has taken more time and more effort than I once imagined it should. But I have come to understand that this is precisely what was needed. The journey of becoming who I was meant to be has not been linear, and that is not a flaw in the path. It is the path. The value has never been in the destination. It was in the going deep, the growing slowly, and the gradually becoming.

For that, too, I am grateful.

ABOUT THE AUTHOR

Ernesto J. Gómez has spent more than three decades inside the trenches of business - first as an entrepreneur, later as a global executive, and now as a trusted advisor to leaders navigating complexity and change.

He began his career building ventures in the food service industry across Mexico and the United States, gaining firsthand experience in the realities of risk, growth, and reinvention. He later transitioned into corporate leadership, serving as Vice President of Human Capital at Grupo ALFA, one of Mexico's largest conglomerates at the time, where he oversaw human capital strategy across 83,000 employees in 32 countries. He went on to become Chief Human Resources Officer of Sigma Alimentos, a $9 billion global consumer packaged goods subsidiary of ALFA, with 45,000 employees across 18 countries.

In these roles, he led large-scale talent and cultural transformations across multiple geographies—experiences that profoundly shaped his understanding of performance, leadership, and the untapped depth of human potential.

Ernesto is the founder and CEO of Aspen Mindset1, an advisory firm dedicated to advancing what he calls **Regenerative Performance**, a new performance logic for the AI era grounded in human capacity, alignment, and meaning.

He is the author of *Regrowth: How Organizations Can Overcome Stalling by Unlocking Their People's Potential*, recognized as an Amazon Hot New Release and honored among the Top 20 Best Business Minds Books of 2025. In *Regrowth*, he introduced the systemic foundations of organizational renewal. In *The Soul Edge*, he turns to a deeper question: What kind of leader must we become when intelligence is abundant, yet wisdom, judgment, and meaning remain profoundly human?

A lifelong learner, Ernesto has completed executive education programs at institutions including Stanford GSB, MIT Sloan School of Management, Wharton, Kellogg School of Management, Harvard Business School, Chicago Booth, London Business School, and IMD Business School. In 2024 and 2025, he was invited to speak at MIT Sloan School of Management in the course *Leading Through Ambiguity*, reflecting his continued engagement with the evolving frontier of leadership.

Beyond titles and roles, Ernesto considers himself a student of the human frontier. His work has been shaped as much by boardrooms as by lived experience, the accountability of delivering results in complex environments, the weight of uncertainty, and the quiet inner work leadership demands.

He has been married for over three decades, is the proud father of three daughters, and a grateful grandfather. Outside of work, he can often be found reading, reflecting, or watching a good soccer match—always curious, still learning.

Learn more at **aspenmindset1.com**

www.ingramcontent.com/pod-product-compliance
Lightning Source LLC
LaVergne TN
LVHW090318210926
848480LV00001B/287

9781959009375